Sheep Feeds and Feeding

in Western Canada

Steve Mason, PhD

©AgroMedia International Inc.
'Science into practice'

Acknowledgements:

A large proportion of this document (used with permission) was originally published by the British Columbia Ministry of Agriculture when the author was Provincial Sheep Specialist and later, Provincial Livestock Nutritionist.

Many thanks to Dale Engstrom, Wray Whitmore and Glenna McGregor for their invaluable comments and suggestions for improvement of the text.

Attributable image credits:

Cover image: licensed iStock photo; Figure 1.1: licensed Alamy stock photo; Figures 1.4, 1.5, 1.8 and 1.9: Norman Criddle in G.H. Clark and M.O. Malte, Fodder and Pasture Plants, Canada Department of Agriculture, 1923; Figure 3.7: Penn State University; Figure 6.1, Condition Scoring Reference, p. 62 and Figure 6.14: UK Agriculture and Horticulture Development Board 2019; Figure 6.6: Canada Plan Service; Figure 6.7; Midwest Plan Service; Figure 6.9: M. McG. Cooper and D.W. Morris, Grass Farming, Farming Press Ltd., 1973.

About the author:

Steve Mason obtained both his B.Sc. (Biochemistry) and Ph.D. (Animal Science) from the University of British Columbia. Following a post-doctoral fellowship in medical pharmacology at the University of Calgary, he became a sheep farmer for several years which led to stints as Provincial Sheep Specialist and later, Provincial Livestock Nutritionist, with the British Columbia Ministry of Agriculture.

Moving back to Alberta, Steve became Manager of ProLivestock Nutrition/Management Specialists, a consulting unit of United Grain Growers, later Agricore United. Subsequently, after a short assignment as Senior Extension Associate with Cornell University's Pro-Dairy program, Steve established AgroMedia International Inc., a business that provides knowledge translation and transfer services as well as contract scientific and technical support to the Canadian livestock industries. Operating under the name 'AgInformatics' the company also provides data management and analysis services to the agricultural research community. More recently, a partnership operating as 'Farm Animal Care Associates' has focused on assisting livestock producers with the adoption of best management practices for animal health and welfare. Since 2010, Steve has served as an Adjunct Associate Professor with the University of Calgary Faculty of Veterinary Medicine, teaching nutrition, mentoring graduate students and participating in research. He is a Registered Professional Animal Scientist and a Diplomate of the American College of Animal Nutrition.

Disclaimer:

Table of Contents

Introduction

The primary goal of any commercial sheep enterprise must be to achieve maximum production at minimum cost. It is the margin between the returns from marketable products and costs of production that determines profitability. Since feed costs usually comprise 60-80% of the total costs of production, it is imperative that feed resources are used to maximize advantage.

The economic realities of sheep production demand a solid understanding of the principles of nutrition leading to the application of sound and profitable management practices. Knowledge of both the nutrient value of available feeds as well as the nutrient requirements of the livestock is essential in order to profitably allocate feed resources.

This guide attempts to put feeding management practices into the whole context of sheep nutrition. Although some of these ideas may seem academic at first glance, they are intended to provide an understanding of the principles which dictate the success (or failure) of management practices. For example, creep feeding works because early consumption of solid feed promotes rumen development.

The target audience for this publication includes sheep producers and their advisors as well as agriculture and veterinary students. Although a few have criticized the technical level of the text as being a little high for sheep producers, the popularity of earlier versions suggests otherwise. In his 40-plus years of involvement in livestock extension, the author has adopted the attitude that it is more appropriate to challenge producers with the scientific basis of animal husbandry that to attempt to oversimplify (dumb-down) the concepts involved. It is hoped that this approach will lead to a more productive, science-based sheep industry in western Canada.

Chapter 1: Feeds

Sheep are herbivores—plant-eaters— so a logical starting point for a discussion on sheep feeding is a description of the plants and plant constituents that they consume.

Although some raise sheep because of their fondness for the animals, in most parts of the world sheep and other ruminant animals are farmed for their ability to utilize an otherwise unusable resource to produce food and fibre suitable for human consumption. As explained in **Chapter 3**, bacteria, fungi and protozoa that inhabit the ruminant forestomach have the unique ability to digest the fibrous parts of plants that are indigestible by mammalian enzymes.

In fact, sheep and other ruminants are relatively inefficient in their conversion of feed to human food. The production of beef and lamb can only be economical when the costs of raw materials are low. For this reason, sheep production is based on forage, either fresh, as pasture, or conserved, in the form of hay or silage. Except in the case of fast growing lambs and ewes nursing multiple lambs, grain and other concentrates should only be viewed as supplements. These feeds are more efficiently utilized either by humans directly or through other animal production systems such as swine and poultry.

Plant composition

The growth of plants

Plant growth is summarized in figure 1.1. All green plants absorb carbon dioxide (CO_2) from the air through their leaves. Water (H_2O), nitrates and minerals are assimilated from the soil by the roots. Sunlight, which is trapped by the green pigment chlorophyll, provides the energy that is used by the plant to make carbohydrates, proteins and other organic constituents from these simple precursors.

Carbohydrates

Carbohydrates are combinations of carbon dioxide and water. The simplest combinations are called sugars and include glucose, galactose, fructose and pentose. These substances are soluble in water and are readily transported through the plant to provide for a variety of requirements. Simple sugars are the building blocks for more complex carbohydrates such as cellulose, hemicellulose, pentosans and starch. Cellulose and hemicellulose are important constituents of the plant cell wall, referred to as structural carbohydrates. Starch, pectin and pentosans are some of the forms in which plants store

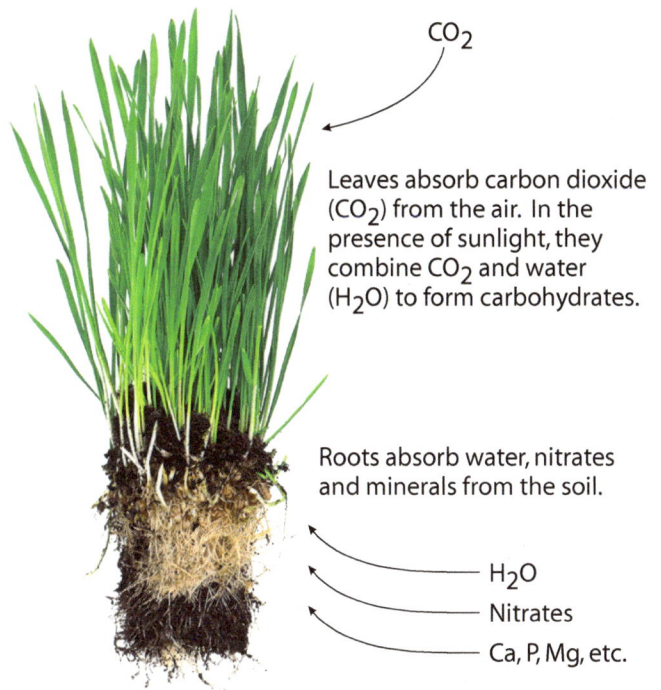

Figure 1.1. Plant growth.

energy. These are often referred to as non-structural carbohydrates, carbohydrate reserves, or in forages like alfalfa, root reserves. Starch is the carbohydrate stored in the endosperm of feed grains.

Fats and oils

Like carbohydrates, fats and oils are made up of carbon, hydrogen and oxygen. However, the proportion of carbon and hydrogen is much greater with the result that fats and oils furnish 2.25 times as much energy per kilogram as do carbohydrates and proteins.

Plants, except for the oilseeds, contain relatively low concentrations of fats and oils. Unlike animals, which store energy as fat, plants store energy as carbohydrate, as suggested above.

Lignin

The more fibrous parts of plants contain considerable amounts of lignin which is a very complex substance, somewhat similar to the complex carbohydrates. It is important in giving plants some of their structural properties, but as we will see later, is of little value as a feed constituent.

Proteins

In addition to carbon (from CO_2), hydrogen (from H_2O) and oxygen (from both), plant proteins contain nitrogen and most contain sulfur and phosphorus assimilated from the soil. These compounds are almost infinite in nature, mainly functioning in plants in the

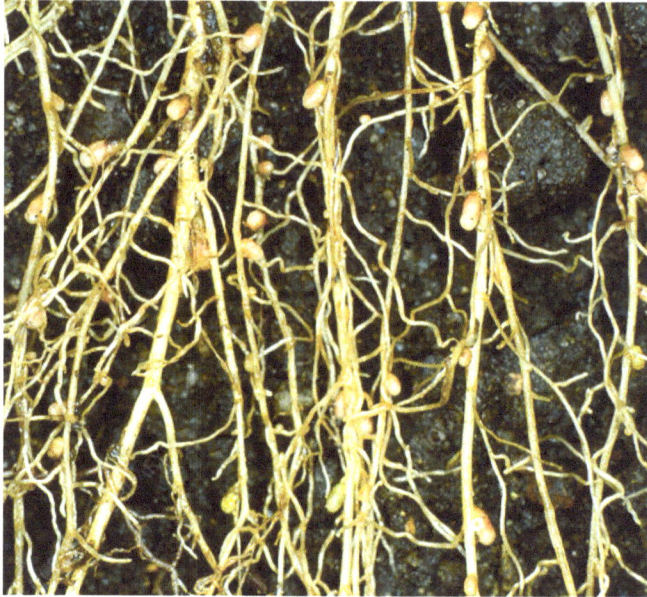

Figure 1.2: Rhizobia root nodules on red clover.

form of enzymes. Proteins are predominantly found in the reproductive parts and in actively growing parts such as leaves. In animals, protein is found mainly in the form of muscle tissue. Of particular importance in sheep production, wool is composed of the protein keratin. Since these constitute the end-products of the sheep enterprise, it is easy to appreciate the importance of protein in the feed.

Legume forages such as alfalfa and clover generally contain higher levels of protein than grasses. This is due to the large supply of nitrogen available to them through fixation from the atmosphere. Nitrogen fixation is facilitated by bacteria (rhizobia) contained in nodules on the roots which transfer nitrates to the plant (figure 1.2). In exchange, the legume plant supplies soluble carbohydrates which provide energy to the bacteria.

Minerals

Plants assimilate minerals from the soil through their roots. Although present in relatively small amounts, they are as essential to the development of the plant as they are to the animals that consume them. Minerals are well distributed in the plant, largely occurring in association with the organic compounds (carbohydrates, proteins, fats and oils). For example, magnesium is an essential component of chlorophyll.

Vitamins

Like the other organic compounds, plants synthesize vitamins from raw materials absorbed through their roots and leaves. Some of them, like the B-vitamins serve much the same functions in plants as they do in animals. Others serve a specific function in

plants which is quite different from their function in animals. For example, carotene is a yellow plant pigment that plays a part in photosynthesis. Animals convert carotene to vitamin A which is essential in maintenance of epithelial (surface) tissues.

Feeds

Forages

Grass and legume forages consist primarily of the vegetative parts (i.e., stems and leaves) of plants containing relatively higher concentrations of structural carbohydrates than the reproductive parts (e.g.., seeds, tubers) that contain higher levels of non-structural carbohydrates and other nutrients.

The nutritional value of forages is largely determined by their stage of maturity as illustrated in figure 1.3. As plants grow and mature, stems and leaves accumulate lignin which is not only indigestible itself, but also forms complexes with structural carbohydrates, inhibiting their digestion. At the same time, the protein content of the plant declines.

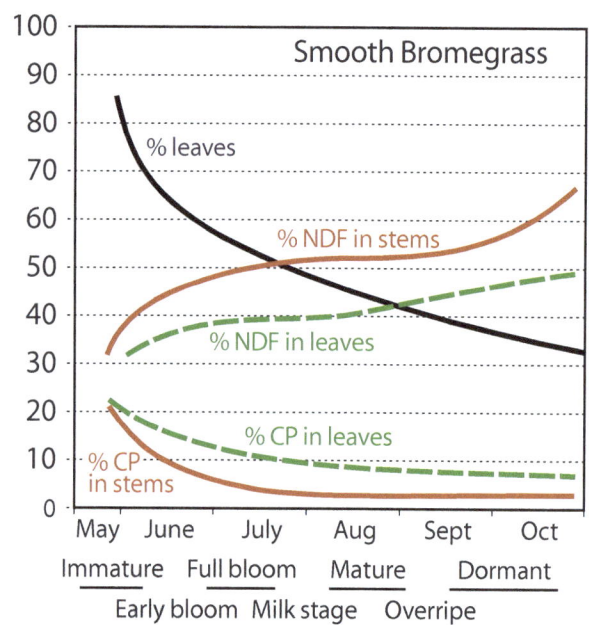

Figure 1.3: Composition changes as a grass crop matures. CP: crude protein. NDF: neutral detergent fibre.

Grass identification
Be aware that, for ease of identification, the following images of various grasses show them at maturity, well beyond the stage at which they would be harvested for feed. As illustrated in figure 1.3, at this stage of growth the nutritional value of the vegetative parts of the plant has reached its lowest level. However, it is often difficult to pecisely differentiate grass species in their earlier vegetative stages without resorting to close inspection of their finer anatomic features.

Figure 1.4a: Perennial grasses at maturity: left - smooth bromegrass; centre - orchardgrass; right - perennial ryegrass.

Perennial grass forages

Perennial grasses commonly used in western Canadian sheep diets include:

Bromegrass: there are two types of bromegrass—a southern type and a northern type. In western Canada, the northern type (smooth bromegrass) is more vigorous and has higher seed yields than the southern type. It is tolerant of drought and extreme temperatures and can be grown alone or mixed with other grasses and legumes. Bromegrass is the most commonly used companion to alfalfa in mixtures grown on dryland.

Orchardgrass is used for pasture, silage and hay. It has a deep, competitive root system and can interfere with nutrient uptake in legumes so the two are not normally grown together, especially when heavily fertilized with nitrogen.

Perennial Ryegrass is rarely grown in the prairie provinces although it is very common in south coastal British Columbia. A short-lived bunch grass with a shallow root system, it is very palatable and nutritious when harvested at the correct maturity.

Tall Fescue can grow even on the poorest of soils. It is tolerant of both acidic and alkaline soils and, although it has a low moisture requirement, tall fescue does especially well under moist conditions. It can also be used to control soil erosion.

Reed Canarygrass is particularly tolerant of low, poorly drained areas. Canarygrass is palatable to cattle as long as it is not allowed to become too mature as it becomes coarse with age. It is often used to stabilize the banks of waterways because of its ability to develop a dense sod.

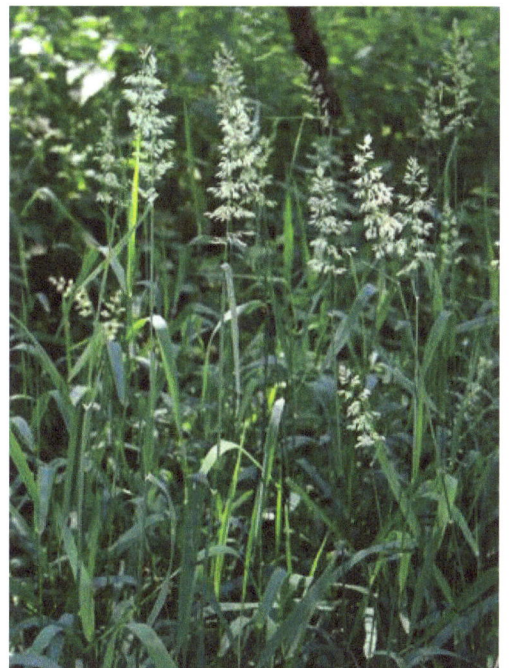

Figure 1.4b: Perennial grasses at maturity: left - tall fescue; right - reed canarygrass.

5

Timothy is used for both pasture and hay; it grows well with legumes; and is easy to harvest. Overmature timothy is very fibrous, limiting intake.

Wheatgrass: there are several different types of wheatgrass: crested, intermediate, pubescent, slender, streambank and tall. These plants are persistent, drought resistant, and can be found throughout the Canadian prairies.

Annual grass forages

Other than the grain crop forages described below, annual grasses are not commonly used in western Canadian livestock diets. The one exception is *Annual (Italian) Ryegrass*, a fast growing, high-yielding species which establishes easily. Although most commonly used for grazing, Italian Ryegrass makes good quality hay or silage when harvested at the correct stage of maturity. Changes in composition with advancing maturity are similar to those for perennial grasses.

Figure 1.4c: Perennial grasses at maturity: left - timothy; right - wheatgrass.

Figure 1.5: Annual (Italian) Ryegrass at maturity

Grain (cereal) crop forages

The 'small' grain crops (members of the grass family *Gramineae*), including barley, rye, oats, triticale and wheat, are common forage sources in western Canadian sheep diets. These are harvested for silage when the grain is in the early development (late milk to soft dough) stage or for dry feed at the 'boot' stage, when the immature seed head emerges from the leaf sheath. When utilized after field drying, these forages may be referred to as greenfeed; when harvested and stored wet, they are often referred to as grain-crop silage.

Barley grows best in well drained, fertile soils. This plant is also adapted to growth in sandy soils. The disadvantage to using barley is the presence of spiky awns on the seed heads which decrease palatability.

Rye is well suited for pasture because of its high productivity. A disadvantage of using rye as pasture is that it quickly becomes unpalatable as it matures.

Oats can be grazed or stored as hay (greenfeed) or silage. Particular varieties (e.g., Foothills) have been bred for use as high quality forages.

Triticale is the result of a cross between wheat and rye. It has the potential to be higher yielding than barley but many attempts to use triticale silage in lactating dairy cow diets have been disappointing due to low intakes.

Wheat can yield quality pasture, greenfeed or silage that is palatable, high in protein and energy potential and low in fibre.

When ensiling small grains the moisture content should be 62–68% to restrict excessive air in the stems and a higher concentration of butyric acid in the silage. As pasture, small grain crops can be grazed during winter months (where weather permits) and into early spring without reducing later harvest yields.

Corn for silage is commonly grown in south coastal British Columbia but often will not achieve optimum

Figure 1.6: Small grain crops should be cut for hay at the boot stage (left); for silage, at the milk to soft dough stage.

maturity when grown on the prairies due to shorter, cooler growing seasons. When harvested for silage, the crop should be cut when it reaches a dry matter level of 30–35% (65–70% moisture). Corn silage is highly palatable and a good source of digestible energy (due to its grain content) but low in digestible protein.

Perennial legume forages

Legume forages are primarily fed to livestock as hay and silage. They are seldom offered as pasture or as fresh-cut forage because of their potential to cause bloat. When ensiled, perennial grasses and legumes are often called hay-crop silages to differentiate them from grain-crop (cereal) silages.

Like grasses, the intake potential and nutritional quality of legumes are largely dependent on their stage of maturity. As the plants mature, their stems elongate, the proportion of leaf to stem decreases, fibre content increases and protein declines. These relationships are illustrated in figure 1.7. In comparison to grasses, legumes are higher in protein, calcium, magnesium, sulphur and copper, but lower in manganese and zinc.

Alfalfa is a perennial legume that can survive in both cold and warm climates. Alfalfa can be used for silages, hay or grazing. This plant cannot withstand drought for long periods of time nor can it tolerate acidic soil or a high water table. Grazing or feeding recently harvested alfalfa can cause bloat in ruminants.

Sainfoin, native to Europe and parts of Russia and Asia, was introduced into western Canada in the 1960s as a non-bloating alternative to alfalfa. It is used to a

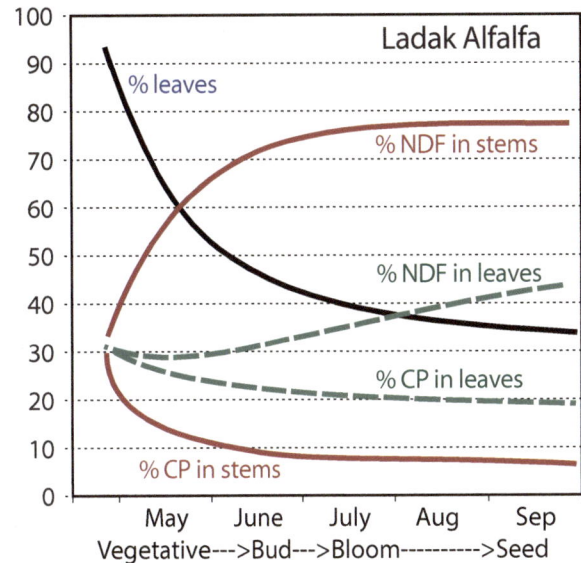

Figure 1.7: Composition changes as alfalfa crop matures. CP: crude protein. NDF: neutral detergent fibre.

limited extent for hay and pasture, although it has poor longevity and poor tolerance to grazing.

Birdsfoot Trefoil is a perennial legume adapted to temperate to cold climates. It is tolerant of drought and acidic soils and can be used for hay or pasture. Trefoil is palatable and nutritious but also has the potential to cause bloat.

Cicer Milkvetch is a perennial that can grow in cold climates. It is tolerant of saline soil, drought, flooding and acidic and alkaline conditions. Because the crop does not cause bloat, it is commonly used for pasture; less frequently for hay.

Figure 1.8: Perennial legumes in bloom: left - alfalfa; centre - sainfoin; right - birdsfoot trefoil.

Red Clover is a perennial that can be grown in moist conditions in cold, warm and acidic soils. It will, however, not tolerate drought. There are two types of red clover: single-cut and double-cut. Single cut varieties flower later, are better able to withstand winter and are larger. Red clover of both types are most suitable for silage; used as pasture there is a risk of bloat.

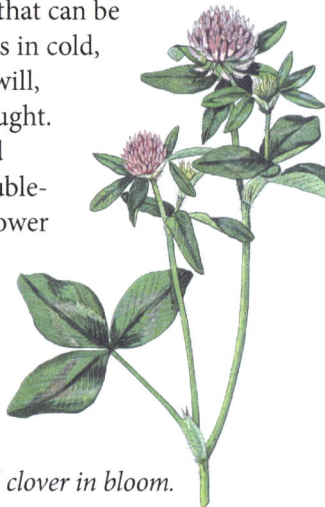

Figure 1.9: Red clover in bloom.

Concentrates

Concentrate ingredients generally contain lower concentrations of fibre relative to those classified as forages. As a result, most concentrates yield higher levels of energy when digested by both animals and intestinal microbes. Concentrates include grains, oilseeds, protein sources, fats and oils, mineral and vitamin supplements, and additives.

With the exception of grains, most small flock owners will not be directly exposed to the other concentrates described below except as they are included in commercial mixed feeds and supplements. Producers managing larger flocks may purchase individual concentrate ingredients as bulk commodities to be blended into complete diets on-farm.

Grains

Grains are the seeds of the cereal crops described above for use as forages, including barley, rye, oats, triticale, wheat and corn. To be efficiently digested by cattle, barley must be processed by either grinding or rolling. This is because the fibrous hull of the whole grain kernel prevents access to its starchy core by both rumen microbes and mammalian digestive enzymes. Conversely, over-processing exposes the starchy endosperm to rapid degradation by rumen microbes—an advantage in terms of maximizing microbial protein synthesis but a disadvantage due to a higher risk of rumen acidosis resulting from a high rate of volatile fatty acid production.

Barley is the principle feed grain used in western Canadian sheep diets. After processing, barley starch is one of the the more rapidly degraded among the common grains, as indicated in table 1.1. However, a number of studies have found that there is no advantage to processing barley to be included in diets for sheep as young as 2 months of age (table 1.2).

Grain Component	Wheat	Barley	Corn
	Degradation Rate, %/hr		
Dry Matter	28.6	17.8	11.5
Starch	34.1	30.9	8.6

Table 1.1: Rumen degradation rates of processed feed grains.

Oats grain contains less starch and more fibre than the other feed grains and, therefore, contributes less energy to the diet than an equal quantity of other grains. At one time, this was considered an advantage—oats grain was fed in preference to other 'hotter' grains because it posed a lower risk of acidosis. With its fibrous hull, oats require processing before feeding but rolling or grinding produces a relatively bulky ingredient.

Rye is similar to barley in its appearance and use as a dietary energy source. It is palatable, but that palatability may decrease with increased inclusion in the diet—for example, it should not make up more

Figure 1.10: Small grains commonly used in western Canadian livestock diets. B - barley; O - oats; R - rye; W - wheat.

| Trial | Variable | ---------- Processing Method ---------- | | | |
		Whole	Ground	Rolled	Pelleted
1	ADG	0.29		0.25	0.22
	F:G	3.85		4.43	3.98
2	ADG	0.23	0.22		
	F:G	5.60	6.26		
3	ADG	0.24	0.24		
	F:G	6.61	6.61		
4	ADG	0.18			0.17
	F:G	7.53			7.76

Table 1.2: Summary of 4 sheep feeding trials comparing performance of growing lambs fed whole barley versus ground, rolled or pelleted barley. ADG - average daily gain (kg/day); F:G - feed to gain ratio (kg/kg).

than 20% of the grain mix. Rye is susceptible to a fungus which produces the mycotoxin, ergot. Grain containing ergot is indicated by the presence of large and misshapen kernels and should not be fed.

Figure 1.11: Ergot on a rye seed head.

Wheat usually has a higher protein content than either barley or corn. Although its energy value is roughly equivalent to that of corn, processed wheat is very rapidly degraded in the rumen (see table 1.1), increasing the risk of acidosis when fed in large quantities (e.g., more than 25% of diet DM). When wheat grown for human consumption fails to meet grade standards, it is often graded as feed wheat which becomes available for livestock feeding at prices competitive with barley and other feed grains.

Corn: Little corn is harvested as grain in western Canada because of inadequate heat during short growing seasons. However, processed corn grain may occassionally be included in diets for growing lambs or lactating ewes, taking advantage of its palatability and its slower rate of rumen degradation (table 1.1). The lower rate of volatile fatty acid production in the rumen lowers the risk of acidosis.

The degradation rates in table 1.1 apply to dry rolled or ground grains. Steam rolling, steam flaking and extrusion increase the starch degradation rates of all grains. This effect is most pronounced for corn as moist heat opens its crystalline starch structure to rumen microbial degradation.

Oilseeds and meals

Canola is widely grown in western Canada for its high quality oil. If they are effectively processed, whole canola seeds can be incorporated into livestock diets. But, unless the right equipment is used, rolling is difficult because of the high (40+%) fat content of the seeds. Finely grooved rollers equipped with scrapers are necessary to keep the oily, crushed product from accumulating on the rollers. The high levels of unsaturated fatty acids in canola oil can reduce fibre digestion by rumen microbes, limiting its inclusion rate to about 10% of diet DM. The total dietary inclusion rate of added fats and oils should not exceed 5% of DM.

Canola meal, a by-product of oil extraction from whole seeds, is widely used as a protein source in livestock diets. In sheep diets, canola meal can be included at up to 10% of diet dry matter, typically as an ingredient in a commercial protein supplement.

Figure 1.12: Canola seed (left) and meal (right).

Soybeans are a heat and moisture-loving crop. Although extensively grown in Ontario and Québec, only in the most southerly (preferably irrigated) parts of the prairie provinces are growing seasons warm enough for routine cultivation. However, if the price is right, whole soybeans can be successfully fed to growing lambs at up to 20% of dietary dry matter.

Soymeal is the product remaining after extracting oil from whole soybeans. It is high in protein and energy and is one of the most commonly used protein supplements in livestock diets. Its palatability makes it particularly suitable as a protein supplement in lamb starter (creep) diets.

Figure 1.13: Soybeans (left) and soymeal (right).

Other protein sources

Peas: five different types of peas that are commonly available in western Canada are shown in figure 1.14. Large and small, green and yellow peas are the most common. Research in North Dakota has demonstrated that field peas can outperform corn in lamb finishing diets and that they are also suitable for feeding to late gestating and lactating ewes. However, because peas and other pulses (e.g., lentils) can cause off-flavours in lamb meat they should not be used beyond a level required to satisfy dietary protein requirements.

Peas are widely variable in their nutrient composition, depending on variety and growing conditions. Samples of 27 pea varieties tested by Alberta Agriculture varied in crude protein content from 23 to 27% of dry matter. In samples from 5 varieties tested at University of Saskatchewan feed lab, starch values ranged from 27 to 50%. Therefore, when including peas in diets, it is important to base inclusion rates on an analysis of each load—'book' values can be very misleading.

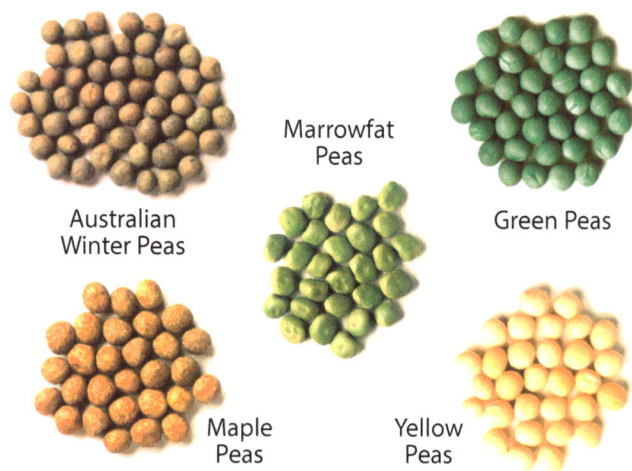

Figure 1.14: Field pea varieties commonly available in western Canada.

Distiller's grains are the by-product from the yeast fermentation of grain to produce ethanol. At the conclusion of the fermentation process, the liquid phase containing the ethanol is separated from the solid grain residue. The dried residue is dried (or dehydrated) distiller's grains (DDG). In most distilleries, after distilling off the ethanol, the remaining liquid is combined with the solid residue and dried. The resulting product is called dried (or dehydrated) distiller's grains with solubles (DDG/S).

Most of the DDG and DDG/S in western Canada is the product of corn distillation for the production of whiskey. This is a high quality product of uniform golden brown colour, containing about 8% fat and 25% crude protein with relatively high bypass (UIP) value. Its palatability and the high digestibility of its fibre (40–45% NDF) make it a valuable ingredient in diets for lactating ewes nursing multiple lambs.

Corn DDG/S imported from the US is usually a by-product of the fuel ethanol industry. Although this product often appears to contain burnt particles that might indicated heat-damaged protein, there is no evidence that it is of inferior feeding quality. Wheat and rye DDG and DDG/S are also sometimes available in western Canada.

If DDGs form a large proportion of the diet, their high phosphorus levels demand calcium supplementation to maintain an adequate dietary Ca:P ratio (see p. 42).

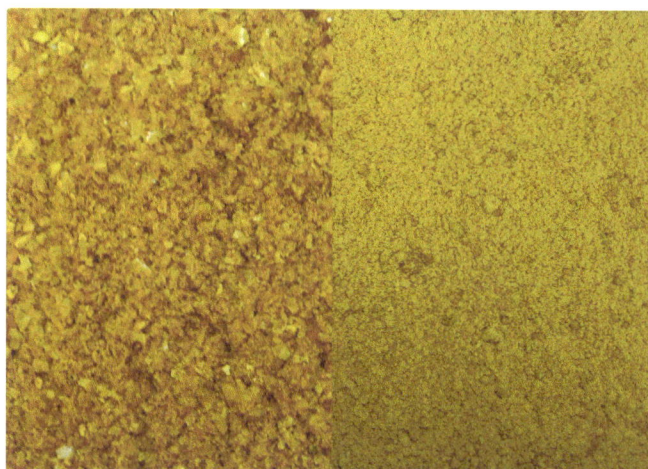

Figure 1.15: Corn distiller's grains (left) and corn gluten meal (right).

Corn gluten meal is a by-product of the manufacture of corn starch and corn syrup. It is often used as a source of bypass protein (UIP) in dairy cow lactation diets but has earned a reputation of being of lower quality due to its low lysine content and inferior digestibility.

Corn gluten meal should not be confused with corn gluten feed which is an entirely different by-product, not readily available in western Canada.

Other plant by-products

Beet pulp is the residue remaining after the extraction of sugar from sugar beets. When molasses is added back to the pulp after processing, the product is called molassed beet pulp or beet pulp with molasses. The addition of molasses increases the product's protein concentration slightly, reduces its fibre content, raises its sugar content and improves its palatability. Beet pulp is commonly available as 'shreds' or pellets.

Figure 1.16: Beet pulp shreds (left) and pellets (right).

Traditionally, dry or moistened beet pulp has been used as a 'top-dress' on forage or other mixed feeds to stimulate intake. Its high fibre content, the high digestibility of that fibre and its palatability make beet pulp a valuable ingredient in ewe lactation diets. This is particularly true when lower quality forages increase the dietary requirement for grain. At these times, beet pulp can be used to replace a portion of the grain, reducing the risk of acidosis and often improving total feed intake. Wet beet pulp has also been used to replace corn silage in diets, serving as a source of both roughage (physical fibre) and energy.

Commercial mixed feeds

The feed ingredients described above are some of the most commonly fed to sheep in western Canada. However, examination of the ingredient list for any of the commercial mixed feeds available in Canada will reveal the presence of many others.

As described in more detail later (p. 14), animal feeds that are imported into or manufactured and sold in Canada are subject to the *Feeds Act and Regulations*, administered by the Canadian Feed Inspection Agency (CFIA). Schedules IV and V of the *Feeds Regulations* lists all feed ingredients approved for use in Canada including detailed ingredient descriptions, standards, labelling and other requirements (figure 1.17).

SCHEDULE IV
(Section 2, subsection 5(2), section 14 and subsections 22(2), 24(3) and 26(8))

PART I

Class 1. Dry Forages and Roughages

1.1 *Alfalfa-grass hay sun-cured ground* (or *Alfalfa-grass meal*) (IFN 1-29-774) consists of the aerial part of a mixture of alfalfa and grass plants (predominantly alfalfa) that has been sun-cured and finely ground. It shall be labelled with guarantees for minimum crude protein, maximum crude fibre, maximum moisture, minimum alfalfa and minimum grass.

1.2 *Alfalfa hay sun-cured ground* (or *Sun-cured alfalfa meal*) (IFN 1-00-111) consists of the aerial part of the alfalfa plant, reasonably free of other crop plants, weeds and mold, that has been sun-cured and finely ground. It shall be labelled with guarantees for minimum crude protein, maximum crude fibre and maximum moisture.

1.3 *Alfalfa leaves meal dehydrated* (or *Alfalfa leaf meal*) (IFN 1-00-137) consists of leaves of alfalfa separated from the alfalfa plant that have been dried by thermal means and finely ground. It shall be reasonably free of other crop plants and weeds. It shall be labelled with guarantees for minimum crude protein, maximum crude fibre and maximum moisture.

Figure 1.17: Example of approved ingredient descriptions from Schedule IV of Canada's Feed Regulations.

Mineral supplements

Virtually all diets for every class of animal require mineral supplementation. For companion animals, swine and poultry, supplementary minerals will normally be incorporated into manufactured feeds containing all dietary requirements. For horses and ruminants, minerals may be included in a concentrate mix and/or may be provided as supplementary minerals mixes, available in several forms:

Complete mineral mixes contain blends of macro- and trace minerals, often with vitamins A, D and E added. Every feed dealer will sell a wide variety of these products, formulated for specific classes of livestock and particular geographic requirements. Commonly sold as blocks or in 10 or 25 kg bags, complete mineral mixes are designed to be offered to animals 'free-choice' (*ad libitum*) or to be blended with other grains and concentrates on-farm.

Figure 1.18: A simple, portable 'free-choice' (ad libitum) mineral feed station providing protection from rain.

SHEEP PASTURE MINERAL
GUARANTEED ANALYSIS

Calcium (Actual) 14.00%
Phosphorus (Actual).. 9.00%
Magnesium (Actual)... 4.00%
Sodium (Actual) 7.00%
Cobalt (Actual) .... 35 mg/kg
Iodine (Actual).. 65 mg/kg
Copper (Added) ....0 mg/kg
Manganese (Actual) .... 2,400 mg/kg
Zinc (Actual)... 6,000 mg/kg
Fluorine (Max.) .... 1,000 mg/kg
Vitamin A (Min.) ....500,000 IU/kg
Vitamin D3 (Min.) .... 50,000 IU/kg
Vitamin E (Min.)... 625 IU/kg

This feed contains added Selenium at 30.0 mg/kg.

INGREDIENTS

Mono-calcium and di-calcium phosphate, calcium carbonate, sodium chloride, magnesium oxide, cobalt carbonate, manganous oxide, zince oxide, potassium iodate, sodium selenite, Vitamin A acetate, Vitamin D3, Vitamin E, canola oil (dust control), Anise-Fenugreek (flavour)

FEEDING DIRECTIONS

Sheep Pasture Mineral is intended to be fed to sheep on a free choice basis while grazing pastures. Sheep Pasture Mineral should be placed in a mineral feeder and there should be 1 mineral feeding station for every 15-20 head of sheep. Expected intake is 7-10 grams per head per day. Sheep should always have free choice access to salt and clean drinking water.

Caution

1. Directions for use must be carefully followed.

2. Do not use in association with another feed containing supplemental selenium.

Figure 1.19: Example of a complete mineral mix for sheep, designed to be offered ad libitum in a pasture mineral feeding station such as that shown in figure 1.18.

Macro premixes are specifically designed for addition to other concentrates in complete concentrate mixes used to supplement home-grown forages or in complete diets including forages (total mixed rations). Containing both macro- and trace minerals as well as vitamins, these premixes are most commonly blended with other ingredients in the feed mill or on-farm.

Micro premixes are used in the same way as macro premixes but they contain only micro- (trace) minerals and vitamins and are, therefore, used in smaller quantities. When micro premixes are used, macro-minerals are usually blended into the complete concentrate mix individually.

Supplementary minerals are most commonly added to diets in the form of inorganic salts. Alternatively, minerals may be provided in the form of chelates— minerals complexed with organic molecules such as amino acids or proteins (referred to as proteinates). Claims are often made that chelated minerals are preferable because they are more bioavailable but the practical relevance of such claims is controversial.

Mineral supplements are either offered *ad libitum* or blended with other ingredients as described above.

SHEEP MINERAL PREMIX
GUARANTEED ANALYSIS

Calcium (actual)... 17.8%
Phosphorus (actual) 4.0%
Magnesium (actual) 7.0%
Sulfur (actual) 0.8%
Sodium (actual).. 10.5%
Potassium (actual) 0.5%
Iodine (actual) .... 60 mg/kg
Iron (actual) .... 1,585 mg/kg
Manganese (actual)................................ 1,680 mg/kg
Zinc (actual) .... 2,800 mg/kg
Cobalt (actual)... 25 mg/kg
Fluorine (maximum) ....212 mg/kg
Vitamin A (minimum) .... 500,000 IU/kg
Vitamin D (minimum)....................... 50,000 IU/kg
Vitamin E (minimum)..........................2,500 IU/kg

Contains no added copper.
This feed contains added Selenium at 30.0 mg/kg.

INGREDIENTS

A list of ingredients used in this feed may be obtained from the manufacturer or registrant.

FEEDING DIRECTIONS

Sheep Mineral Premix is designed to provide sheep with a high level of vitamins, minerals and salt while feeding in a dry lot. Mix with on-farm grains to provide 15 grams of premix to young ewe lambs, 25 grams of premix to dry ewes and 33 grams of premix to lactating ewes.

Figure 1.20: Example of a macromineral premix designed to be blended with other ingredients on-farm.

Figure 1.21: Providing iodized or cobalt-iodized salt licks for sheep on pasture is common, but do not expect them to satisfy their essential trace mineral requirements from trace mineral licks.

Unless there is no other practical choice, *ad libitum* feeding is not recommended except for iodized or cobalt-iodized salt. Although the rate of disappearance of mineral from an *ad libitum* feeder, block or 'lick' may seem appropriate for a group of animals, product loss from weather and spillage will almost always occur and individual intakes will vary widely. An often-heard argument in favour of *ad libitum* feeding of minerals is that animals have 'nutritional intelligence'—that they know what they need. This is no more true for sheep than it is for humans.

Vitamin supplements

Supplementary vitamins are added to mixed feeds for all classes of domestic animals. For sheep, vitamins may be included in a concentrate mix and/or may be provided with supplementary mineral mixes, as described above. Although most forages and grains contain substantial quantities of vitamins A, D and E at harvest, by the time these feeds are consumed, vitamin levels may have dropped considerably due to the effects of sunlight, rain damage, heat, oxidation and mould growth.

Because rumen microbes can synthesize many of the essential B-vitamins, mixed feeds for ruminants usually provide only the fat soluble vitamins: A, D and E. In most cases, these provide adequate intakes of A and D. However, unless specifically formulated for the requirements of high production (rapidly growing lambs, ewes nursing multiple lambs), the vitamin E levels in these mixes are generally inadequate, vitamin E being added only for the purpose of reducing oxidation of other ingredients.

Vitamins may also be delivered by intramuscular injection, as an alternative to feeding. The injection route is particularly appropriate for animals who are not consuming mixed concentrates, such as ewes in late gestation or animals on pasture.

Anti-nutritive factors in feed

Mycotoxins

Fungal infestation (mould growth) can occur at any time during plant growth, harvesting, storage or processing, primarily influenced by moisture level, temperature, and availability of oxygen. Not all moulds that grow on plants produce toxins and the mycotoxins that are produced by different moulds vary in their effects on the animals that consume them.

Mycotoxin	Effect on animals
Aflatoxin	liver disease, carcinogenic and teratogenic effects
Trichothecenes	immunologic effects, hematological changes, digestive diorders, edema
Zearalenone	estrogenic effects, atrophy of ovaries and testicles, abortion
Ochratoxin	nephrotoxicity, mild liver damage, immune suppression
Fumonisin	pulmonary edema, hepatotoxicity, leukoencephalomalacia, nephrotoxicity,
Ergotamine	vasoconstriction, necrosis of the extremities, agalactia, abortion
Vomitoxin	feed refusal, poor weight gains negative health effects are rare

Table 1.3: Mycotoxins commonly found in animal feeds and their effects on animal health.

Nitrates

As described earlier, nitrate is the form of nitrogen the plant roots take up from the soil, from which it is transported to the leaves. When plants are stressed, excess nitrates may accumulate. Drought or hot, dry winds put forage under water stress, often resulting in nitrate accumulation. Cool, cloudy weather or crop damage caused by hail or frost impairs photosynthesis, resulting in higher nitrate levels. When any of these conditions occur within a few days of harvest or grazing, the potential for nitrate poisoning exists. If the stress is removed and the plants recover, nitrate levels will return to normal within several days.

In ruminant animals, nitrate is converted to nitrite and subsequently ammonia by rumen microbes. Nitrate poisoning occurs when the nitrite level in the rumen exceeds the capacity of the microbes to convert it to ammonia. When this happens, both nitrate and nitrite are absorbed into the bloodstream. Nitrite combines with hemoglobin to form methemoglobin, reducing the oxygen carrying capacity of the blood, ultimately leading to tissue oxygen starvation.

Listeria

Mouldy or spoiled silage that has not reached pH 4.0–4.5 during fermentation may harbour *Listeria monoctyogenes*, bacteria that normally dwell in soil and in the gastrontestinal tract. Sheep consuming contaminated silage can develop Listeriosis, commonly referred to as 'circling disease' resulting from infection of the central nervous system. Contamination is most commonly due to soil incorporation during harvest.

Signs of infection may include depression; fever (40–41°C); weakness; incoordination, circling, nasal discharge, loss of appetite or facial paralysis—drooping eyelid or ear on one side. The neck may flex away from the affected side; the animal may lean or push on stationary objects. Pregnant ewes may experience placentitis, fetal infection and death, abortion, stillbirths, neonatal deaths and metritis.

Feed regulations

In Canada, the manufacture, sale and import of animal feeds are regulated by law under the *Feeds Act and Regulations*, administered by the Canadian Food Inspection Agency (CFIA). All feeds must be safe for the animals to which they are fed, to those handling or otherwise exposed to the feeds, and to the environment. They must also be deemed safe to humans through the potential transfer of residues in meat, milk and eggs.

Feeds must also be shown to be effective for their intended purpose. As mentioned earlier (p. 11), feed ingredients acceptable for manufacture, import or sale as livestock feed have been reviewed and assessed by CFIA and are referenced in the *Feeds Regulations* along with appropriate guarantees, standards and labelling requirements. All imported feeds must meet the same standards as domestic feeds.

Feed labelling

According to the *Feeds Regulations* all livestock feed (bagged or bulk) manufactured, sold or imported must be properly labelled. In the case of bulk feed, a label is required to accompany the shipment while labels must be affixed to the package for bagged feeds.

Specific rules apply to the contents of feed labels for each type of medicated or non-medicated feed (commercial, customer formula, consultant formula, veterinary prescription) and for the items listed in the guaranteed analysis.

Feed additives

Feed additives are defined as non-nutritive substances added to feeds to:
- improve the efficiency of feed utilization;
- stimulate growth or other types of production;
- increase feed acceptance;
- enhance food safety and stability;
- improve the health or metabolism of the animal.

In Canada, any feed additive that is classified as a 'medicating ingredient' (e.g., antibiotic, dewormer, growth promotant, β-agonist) is regulated by the *Feeds Act* and *Regulations*, administered by the Canadian Food Inspection Agency (CFIA). The CFIA's Compendium of Medicating Ingredient Brochures (CMIB) lists those medicating ingredients permitted by Canadian regulation to be added to livestock feed. This includes drug products that may only be used under a veterinarian prescription as well as products that may be used in the manufacture of livestock feed without veterinarian approval (over-the-counter (OTC) products). The CMIB specifies:
- the species of livestock;
- the level of medication;
- directions for feeding;
- the purposes for which each medicating ingredient may legally be used (claims);
- the brand of each medicating ingredient that is approved for use in Canada;
- labelling requirements.

Table 1.4 is a facsimile of the CMIB page that lists the medicating ingredients allowed as additives to sheep feeds in Canada. Table 1.5 is an example of the detail provided in the CMIB for a medicating ingredient.

In addition to compliance with the *Feeds Act and Regulations*, medicated feeds must also comply with the *Food and Drugs Act and Regulations* administered by Health Canada, as follows:
- In keeping with the *Feeds Regulations*, all medicated livestock feed imported, manufactured or sold in Canada must meet standards set out in the CMIB unless the feed is a veterinary prescription feed (a feed manufactured pursuant to a veterinary prescription).
- A medicated feed is exempt from complying with the standards set out in the CMIB only if the medicated feed has been manufactured pursuant to a veterinary prescription (see below) and if, amongst other criteria, the source of the medicating ingredient prescribed and used in the medicated feed complies with the *Feeds Regulations* and conditions set out in the *Food and Drug Regulations*.

Species (class)	Name of medicating ingredient	MIB code	Name of approved brand(s)	Status
Lambs (fed in confinement)	Chlortetracycline hydrochloride	CTC	1. Aureomycin 220 G Granular Medicated Premix 2. Chlor 100 Granular Medicated Premix	Veterinary prescription
Lambs (fed in confinement)	Decoquinate	DEC	Deccox 6% Premix	Over the counter
Lambs (fed in confinement)	Lasalocid sodium	LAS	1. Avatec 20 Lasalocid Sodium Premix 2. Bovatec 20 Lasalocid Sodium Premix	Veterinary prescription
Lambs (fed in confinement)	Oxytetracycline hydrochloride	OTC	1. Terramycin – 50 2. Terramycin – 100 3. Terramycin – 200 4. Oxysol – 110 5. Oxysol – 220 6. Oxysol – 440 7. Oxytetracycline 50 Granular Premix 8. Oxytetracycline 100 Granular Premix 9. Oxytetracycline 200 Granular Premix	Over the counter
Sheep	Monensin	MOS	Coban Premix Rumensin Premix	Over the counter

Table 1.4: Medicating ingredients approved for sheep - July 2020.

Decoquinate (DEC) – Medicating Ingredient Brochure

Lambs

Table of approved claims and brands

Approved claims	Name of approved brand(s)	Drug concentration in DIN product
Claim 1	Deccox 6% Premix	Decoquinate at 60 g/kg

Claim 1
As an aid in the prevention of coccidiosis caused by pathogenic Eimeria spp. in lambs fed in confinement.

Level of drug in complete feeds or complete diet
At a level which will provide 0.5 mg decoquinate per kg of body weight.

Directions for use in a complete feed
Feed for at least 28 days during periods of coccidiosis exposure, or when experience indicates that coccidiosis is likely to be a hazard.

Warning
No withdrawal period is required when this medicated feed is fed at the recommended rate of 0.5 mg decoquinate per kg of body weight per day.
Do not use in lactating ewes producing milk for human consumption.
Keep out of reach of children. (Required on premix and supplement labels only.)

Caution
This product should only be used in lambs infected, or likely to become infected, with Eimeria spp.
The use of decoquinate will maintain normal growth under conditions of coccidial challenge but does not improve growth of healthy lambs.
Efficacy of decoquinate treatment during clinical outbreaks of coccidiosis in lambs has not been demonstrated.

Accepted Compatibilities
Nil

Table 1.5: Medicating Ingredient Brochure details for Decoquinate.

Health Canada's Drug Product Database (DPD) contains product specific information on pharmaceuticals and disinfectants that are currently approved for oral administration to sheep. An example is shown in table 1.6.

Current status:	Marketed
Current status date:	1997-11-18
Original market date:	1984-12-31
Product name:	IVOMEC
Description:	IVOMEC DRENCH FOR SHEEP
DIN:	00622125
Product Monograph/ Veterinary Labelling Date:	2017-11-01
Company:	MERIAL CANADA INC
Class:	Veterinary
Species:	Sheep
Dosage form(s):	Solution
Route(s) of administration:	Oral
Number of active ingredient(s):	1
Schedule(s):	OTC
American Hospital Formulary Service (AHFS):	Not Applicable
Anatomical Therapeutic Chemical (ATC):	Not Applicable
Active ingredient group (AIG) number:	0124785011

List of active ingredient(s)

Active ingredient(s)	Strength
IVERMECTIN	0.8 MG / ML

Table 1.6: An example entry in Health Canada's DPD.

Veterinary prescription feeds

The *Food and Drug Regulations* state that a person may sell a medicated feed pursuant to a written prescription of a veterinary practitioner if:

- all drugs used in the medicated feed as medicating ingredients have been approved for sale by Health Canada and each drug has either:
  - a valid DIN (drug identification number), or;
  - been permitted for sale as an Investigational New Drug (IND), an Emergency Drug Release (EDR) or an Experimental Studies Certificate (ESC).
  - the medicated feed is for the treatment of animals under the direct care of a veterinary practitioner who signed the prescription.

Off-label use of drugs

The term 'off-label' refers to any use of a drug that varies from the permitted uses specified in Health Canada's Drug Product Database as indicated on the product label. Usually, this means using a drug for an illness or disease other than the authorized reasons for use—an 'off-label indication'. Modified dosages, administration frequencies, and the type of patient being treated (species and age) may also be considered 'off-label'.

Off-label drug use is permitted only when prescribed by a veterinarian.

Antimicrobials as feed additives

In the past, antibiotics were permitted as feed additives in Canadian livestock diets for the purpose of enhancing feed efficiency and growth. This practice is being phased out with exception of a class of antibiotics called ionophores which includes lasalocid and monensin. While ionophores are technically classified as antibiotics, they are not used in human medicine and are, therefore, not of concern for contributing to the development of antibiotic resistance in the human population.

Chapter 2: Feed Sampling and Analysis

Many of the forages grown in western Canada are of insufficient quality to meet the minimum requirements of our livestock without supplementation. For example, a significant proportion of our grass hays have crude protein (CP) levels in the 4–7% range while 9% CP is considered minimal for ewe maintenance.

Unless the nutrient content of dietary ingredients is known, their allocation to meet productive requirements is pure conjecture. Undernutrition is detrimental to productivity while overnutrition is a waste of valuable feed resources. Maximum dollar returns demand that animal requirements be matched by nutrient intake. This can only be accomplished when nutrient levels in feed have been determined through feed testing.

Feed sampling

Feed testing involves both sampling and analysis. Often the importance of the former is underestimated. A feed testing laboratory can only analyze what is submitted and unless the sample is representative of the available feed supply, the time and money spent will be wasted.

For example, a bale is thrown from the top of a haystack, the strings are cut and a handful of hay is pulled from the centre, put in a plastic bag and sent in for analysis. Several sources of sampling error are possible:

- Perhaps the bale, being on top of the stack, was one of the last baled. Did that part of the swath receive more rain than that baled earlier. Had it dried more, resulting in greater leaf shattering? Was it from a corner of a field not typical of the rest?
- As a result of being on top of the stack was the sample bale more weathered? Had the stack received rain?
- In pulling hay from the bale, were leaves stripped off resulting in a sample which had a high proportion of stems?

The goal of feed sampling is to obtain a sample which is representative of the average feed quality of the bulk of feed from which it is taken. If there is reason to believe that significant differences exist between one batch of feed and another, then representative samples should be drawn from each batch separately.

Sampling hay

A convenient coring tool (figure 2.1) is available for sampling hay. A stack should be sampled in 10 to 15 different locations, well distributed around the stack. Additional cores will result in a more representative sample. The cores should be thoroughly mixed in a plastic bucket and from this a sub-sample should be taken for submission.

Figure 2.1: A coring tool used for sampling hay.

The ideal method for obtaining a sub-sample is to dump the combined cores after mixing onto a flat, smooth surface to form a uniform pile. Then, using a large flat blade, divide the pile into quarters. If a quarter is too large to submit, use the blade to divide the quarter into halves or quarters again, until sub-samples of appropriate size are obtained. A second sub-sample should be saved in case a question arises out of the analysis results.

Sampling silage

Silage in bunkers, piles or bags can be sampled using an extended coring tool such as that shown in figure 2.2. If an appropriate probe is not available, or for silage stored in upright silos, the sampling strategies described in the text box on page 18 are recommended.

Figure 2.2: An extended probe used for sampling silage.

Silage sampling strategies

Silage samples should be obtained both as the silo is being filled and after the silage has been allowed to reach its final stage of fermentation:

- sample(s) obtained when filling the silo will give a good indication of forage quality early enough to properly plan the feeding program;
- samples taken when the silo is opened up will provide additional information on the success of the ensiling process and confirm the pre-ensiling analysis results.

Samples of wet, fresh forage pre-ensiling should be taken as follows:

1. Have a clean, 5 gallon pail with a sealable lid located at the silo.
2. The person filling the silo should take a handful of fresh forage from every 4 loads and place in the pail. Put the lid on the pail after each sample is added to retain the moisture.
3. At the end of each day, mix the fresh forage in the pail by hand and, taking a single handful of the forage, put it in a sealed bag in a freezer. Repeat this process each day for up to 7 days.
4. When finished filling that silo or those silo bags, add the samples together and submit for analysis. Re-freeze the samples prior to shipping if necessary.
5. Separate samples should be taken and submitted if filling silo or silo bags takes longer than 7 days or if field or crop conditions change, for example, if a variety of corn silage is later maturing and therefore greener and wetter than other fields.

Samples of fully-fermented silage from a silo or bag should be taken as follows:

Pit or bunker silos:

1. Don't take the sample until the silo has been opened up and used for at least a week.
2. Take the sample at some point in the day after some silage has been removed. Avoid collecting any silage that has spoiled or has remained exposed for a day or more.

3. Take 20 handfuls of silage from all across the face of the silo going as high as you can reach. Put these into a clean pail and mix thoroughly.
4. Put a sub-sample (2–3 handfuls) of the silage collected in the pail into a plastic bag, freeze and submit to the lab.

Silage bags:

1. Don't take the sample until the bag has been opened up and used for at least 2 days.
2. Take 5 handfuls of silage from the face of the bag and place in a plastic bag in a freezer. Repeat for 5 days.
3. At the end of the 5 days combine the 5 samples and mix thoroughly. Put a sub-sample (2–3 handfuls) in a bag and submit to the lab. Refreeze if necessary

Tower silos:

1. Don't take the sample until the silo bag has been opened up and used for at least 5 days.
2. Take 5 handfuls of silage while the silo is unloading and place in a plastic bag in a freezer. Repeat for 5 days.
3. At the end of the 5 days combine the samples and mix thoroughly. Submit a sub-sample (2–3 handfuls) to a lab for testing. Refreeze if necessary.

NOTE: preserving the actual moisture content in the fresh forage or silage sample is very important. Always double bag silage samples. Freezing or refrigeration and exclusion of air from the sample are necessary to prevent spoilage during transportation to the lab. Send samples to the lab early in the week (Monday preferred) so that they don't sit over a weekend in a warm receiving area of the lab.

Packaging and labelling

Label the sample clearly with an identification description, method or system. You must be able to locate the feed when the results come back if you are to realize the value of the feed test. For example:

Farm	Feed	Cut/variety	Date Collected
Woolly Acres	Corn Silage	Maxim	29 Sept 2017
Footrot Flats	Alfalfa Silage	1st/Regal	5 Jul 2017

source: Dale Engstrom

Sampling concentrates

Dry concentrates stored in bulk can be sampled by plunging an arm into the feed and sampling at least 10 different sites. If the feed is bagged, samples should be taken from five to ten individual bags, ideally using a sampling tool such as one of those shown in figure 2.3.

Nutrient concentrations in grains are much less variable than those in forages. In many cases, such as when grain is purchased in small loads, 'book values' are used. Examples can be found in Appendix A. On the other hand, nutrient levels in screenings and other by-products are impossible to predict. They should be sampled and submitted for analysis.

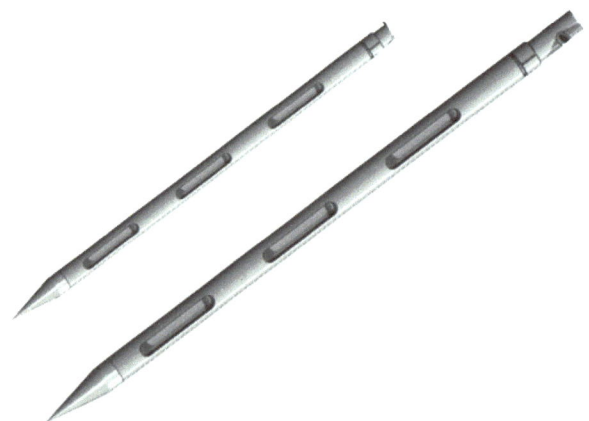

Figure 2.3: Bagged concentrate sampling probes (triers).

Feed analysis

Chemical analysis of feed is aimed at estimating its nutritional value. A brief description of the methods used will provide an understanding of how the results can be used to best advantage. Reference to the feed analysis report shown in figure 2.4 (p. 20) will assist in putting the discussion into perspective.

Dry Matter and Moisture (line 1)

The amount of moisture contained in feeds is widely variable. Hay and grain usually contain about 10% moisture. Silage may contain 50–75%. Pasture plants are often 80–85% water. In most (but not all) feeding situations, animal intake is limited only by the dry matter content of a feed. In other words, a ewe capable of consuming 2 kg (4.4 lbs) of leafy grass hay (10% moisture, 90% dry matter) will also be capable of consuming 9 kg (19.8 lbs) of leafy grass pasture (80% moisture, 20% dry matter). In both cases she will consume 1.8 kg (4 lbs) of dry matter. Expressing feed analysis, animal intake and nutrient requirements on a dry matter basis eliminates moisture as a variable in the comparison of different feeds and in the formulation of balanced rations.

Dry matter content (DM%) may also provide information about the storage properties of feed. Extra moisture may result in heating and spoilage in hay and grain. Inadequate moisture in silage may result in poor preservation while a high moisture content may lead to excessive nutrient leaching. However, for these purposes, dry matter content should be measured before the feed is stored, commonly done using an on-farm microwave oven (see sidebar on page 25). Analysis results from the feed lab are usually received too late for remedial action to be taken.

On the feed analysis report, dry matter is expressed as a percentage. The figure is derived by simply weighing a sample of feed before and after drying at 70–80°C:

$$DM\% = \frac{dry\ weight}{wet\ weight}\ x\ 100$$

Guaranteed analysis values on commercial feed labels are stated on an 'as-fed' (AF) basis. These can be converted to the dry matter basis as follows:

$$DM\ value = \frac{AF\ value}{DM\%}\ x\ 100$$

Acid Detergent Fibre (ADF; line 2)

For the determination of Acid Detergent Fibre (ADF), feed samples are boiled in a solution containing sulfuric acid and the detergent, cetyl trimethyl ammonium bromide. Hemicelluloses and cell wall proteins are dissolved, with the residue containing cellulose, lignin, lignified nitrogen, cutin, silica and some pectins. ADF% is simply the weight of the residue expressed as a percentage of the original sample. As described in the **Energy** section below (p. 24), ADF% is commonly used to estimate the potential of feed to provide energy.

Neutral Detergent Fibre (NDF; line 2)

Neutral Detergent Fibre (NDF) is determined by boiling a sample of feed in a solution containing sodium lauryl sulfate. This detergent extracts lipids, sugars, organic acids and other water soluble components as well as pectin, Non-protein Nitrogen (NPN) compounds, soluble protein and some of the silica and tannin. NDF is the insoluble residue made up of cellulose, hemicellulose, lignin, lignified nitrogen, some protein, minerals and cutin. NDF% is the weight of the residue expressed as a percentage of the original sample. Since it provides the most complete measure of cell wall components, NDF is used to balance fibre requirements in diet formulation.

Crude Fibre (CF; line 3)

Crude Fibre is one of the analytical fractions measured in the proximate system of analysis described below. Although seldom used today in diet formulation, commercial feed tags are still required by law to indicate the Crude Fibre content of the product (see figure 2.6).

The Proximate System of Analysis

The proximate system for routine analysis of animal feedstuffs was devised in the mid-nineteenth century at the Weende Experiment Station in Germany. It was developed to provide a very broad classification of feed components. The system consists of the analytical determinations of dry matter, crude protein, crude fat (ether extract), crude fibre and ash. Nitrogen-free extract (NFE), more or less representing sugars and starches, is calculated by difference rather than measured by analysis:

% NFE = 100% – (CF% + CP% + EE% + Ash%)

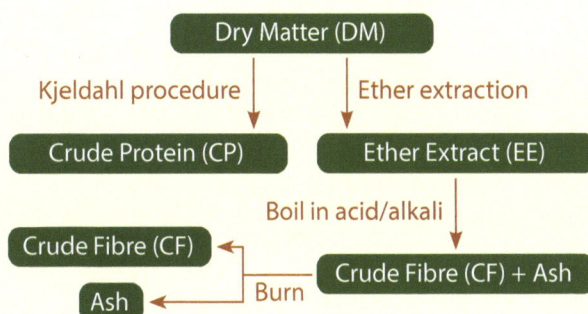

AgriLabs
International Inc.

Feed · Soil · Water

#167, 386 Lower Main Street
Bigtown BC V3W 4X6 Canada
(250)542-6111 · fax: (250) 542-6112
info@ agrilabs.ca · www.agrilabs.ca

Submitted by:

```
Joe Shepherd
Com 1 Site 3 RR#2
Upper Cutbank BC V2Z 0X0
```

Feed Analysis Report

Date received: 17 Aug 2020
Date reported: 21 Aug 2020
Sample retained: 28 Aug 2020
Sample received via: Courier

Lot number: 123456
Report number: 789101
Sample ID: 12131415
Page: 1 of 1

Sample description:

```
second cut
bromegrass hay
north quarter
baled 5 August 2020
```

L	Analysis	Units	Result	Analysis	Units	Result
1	Dry Matter	%	88.3	Moisture	%	11.7
2	Acid Detergent Fibre	% DM	40.3	Neutral Detergent Fibre	% DM	65.9
3	Crude Fibre	% DM	34.6	Crude Fat (EE)	% DM	1.64
4	Crude Protein	% DM	8.34	Non-protein Nitrogen	% DM	1.12
5	True Protein	% DM	7.22	Soluble Crude Protein	% CP	24.6
6	ADF Nitrogen	% DM	0.23	NDF Nitrogen	% DM	0.73
7	Starch	% DM	2.64	Ash	% DM	8.84
8	Calcium	% DM	0.55	Phosphorus	% DM	0.18
9	Magnesium	% DM	0.18	Potassium	% DM	1.59
10	Sodium	% DM	0.04	Chloride	% DM	0.19
11	Iron	mg/kg	147	Manganese	mg/kg	67.4
12	Zinc	mg/kg	16.2	Copper	mg/kg	5.96
13	Selenium	ug/kg	116	Molybdenum	mg/kg	2.23

L	Calculated Value	Units	Result	Calculated Value	Units	Result
14	Non-fibre Carbohydrates	% DM	15.3	Non-structural Carbohyd	% DM	19.8
15	Adjusted Crude Protein	% DM	8.34	Relative Feed Value	%	81.1
16	Digestible Protein	% DM	4.51	Metabolizable Protein	% DM	3.15
17	Degradable Intake Prot	% CP	65	Undegradable Intake Prot	% CP	35
18	Total Dig Nutrients	% DM	58	Digestible Energy	Mcal/kg	2.55
19	Metabolizable Energy	Mcal/kg	2.09	Net Energy, maintenance	Mcal/kg	1.27
20	Net Energy, gain	Mcal/kg	0.73	Net Energy, lactation	Mcal/kg	1.27

Figure 2.4: Example feed analysis report.

A sample (weight A) of the feed residue remaining after extraction of Crude Fat (see EE method below) is sequentially boiled in sulfuric acid and sodium hydroxide. The remaining residue is washed with hydrochloric acid, followed by petroleum ether, dried at 105°C and weighed (weight B). This residue is burned in a furnace at 550°C, and the resulting ash is weighed (weight C).

$$\text{Crude Fibre \%} = 100 \times \frac{B - C}{A}$$

Crude Fat (EE; line 3)

Lipids (fats and oils) are reported in the proximate analysis scheme as Crude Fat, commonly referred to as Ether Extract (EE). A dried sample of feed is placed in a porous extraction thimble in a Soxhlet extractor apparatus (figure 2.5) and subjected to a continuous extraction with diethyl ether at 40–60°C for a defined period of time. After evaporation of the solvent, the residue will contain lipids (fats and oils), waxes, organic acids, alcohol and pigments.

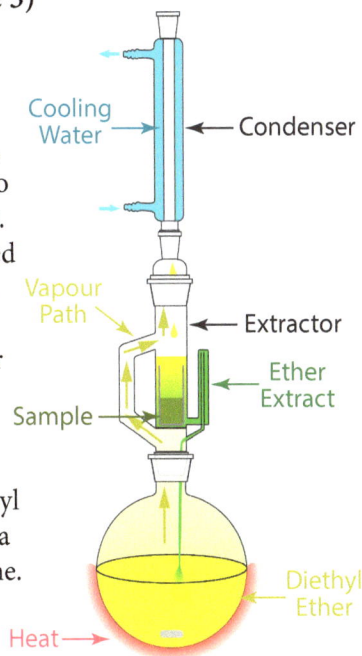

Figure 2.5: Soxhlet extractor used for crude fat analysis.

After extraction, a sample of the defatted feed residue remaining in the thimble is used for Crude Fibre analysis, as described previously.

Crude Protein (CP; line 4)

Determined using the Kjeldahl procedure, Crude Protein is another of the fractions in the proximate analysis scheme (p. 19). A dried sample is first digested in concentrated sulfuric acid with a mercury or selenium catalyst, which converts most of the nitrogen (N) to ammonium sulfate (N present as nitrate is only partially converted). This mixture is cooled, diluted with water and neutralized using sodium hydroxide, resulting in the dissociation of ammonium sulphate. Distillation drives off ammonia and the distillate is titrated with acid to determine its ammonium concentration, from which the N level in the original sample is calculated.

36% SHEEP SUPPLEMENT

GUARANTEED ANALYSIS

Crude Protein (minimum)................................36.0%
Crude Fat (minimum) 3.2%
Crude Fibre (maximum).....................................5.0%
Calcium (actual)...3.5%
Phosphorus (actual)..0.8%
Magnesium (actual)...1.0%
Sulfur (actual)...0.4%
Sodium (actual)..0.84%
Potassium (actual) 1.7%
Iodine (actual)..8 mg/kg
Iron (actual)...231 mg/kg
Copper (actual) .... 15 mg/kg
Manganese (actual).................................355 mg/kg
Zinc (actual)..430 mg/kg
Cobalt (actual)...3 mg/kg
Vitamin A (minimum) ....37,000 IU/kg
Vitamin D (minimum)..........................6,200 IU/kg
Vitamin E (minimum)............................ 260 IU/kg

This feed contains added Selenium at 2.0 mg/kg.

INGREDIENTS

A list of ingredients used in this feed may be obtained from the manufacturer or registrant.

FEEDING DIRECTIONS

For mature sheep – mix with grain and feed as to provide 0.35 kg of this product per head per day.

For growing lambs – mix with grain and feed as to provide 0.15 kg of this product per head per day.

Figure 2.6: Feed labels for manufactured complete feeds as well as supplements and macro premixes must include crude protein, crude fat and crude fibre guarantees.

Since most feed proteins contain about 16% N, CP% is estimated by multiplying the N concentration in the feed by 6.25—the inverse of 16% (1 ÷ 0.16 = 6.25). However, some portion of the N in most feeds is found as Non-protein Nitrogen (NPN) (see below) and, therefore, the value calculated by multiplying N x 6.25 is referred to as Crude rather than True Protein.

Non-Protein Nitrogen (NPN; line 4)

As noted above, Crude Protein is composed of two nitrogen-containing feed fractions: True Protein and Non-protein Nitrogen (NPN). To separate these two fractions, **True Protein (TP; line 5)** is precipitated out of solution using either tungstic acid or trichloroacetic acid, then dried and weighed. The NPN left in solution is measured using the Kjeldahl procedure as described for the determination of CP.

NPN, derived from compounds such as urea, ammonium salts, amino acids, small peptides and nucleic acids, represents the fraction of Crude Protein most rapidly degradable by microbes in the rumen.

Soluble Crude Protein (SCP; line 5)

The Soluble Crude Protein (SCP) content of a feed is estimated by mixing a sample in borate-phosphate buffer. Crude Protein (CP) which is soluble in the buffer is measured using the Kjeldahl procedure. SCP is an alternative to NPN to estimate the amount of CP which is rapidly degradable by rumen microbes.

ADF Nitrogen (ADF-N; line 6)

When silage is put up too dry (greater than 50% dry matter) or hay too wet (less than 85% dry matter), excessive heating may cause some of the protein in the crop to become irreversibly bound to lignin. Heating during the processing of feeds (e.g., pelleting) can have the same effect.

The severity of heat damage is estimated in the feed lab as ADF Nitrogen by measuring the amount of nitrogen (N) associated with the Acid Detergent Fibre (ADF) residue using the Kjeldahl method. ADF Nitrogen may be reported as:
- Acid Detergent Insoluble Nitrogen (ADIN);
- Acid Detergent Insoluble Protein (ADIP);
- Acid Detergent Fibre Protein (ADF-P), or;
- Heat-damaged protein;

expressed as a percentage of either total N, total Crude Protein or total Dry Matter. Nitrogen values are multiplied by 6.25 to convert to protein values.

Although ADF-N is widely accepted as a good measure of heat damage in forages, it may not be appropriate for use in non-forage protein sources such as dried distillers grains (DDG) where there may be very little correlation between ADIN and protein digestibility. In fact, with some non-forage protein supplements, heating may increase their value as Undegradable Intake Protein (UIP; see p. 23).

NDF Nitrogen (NDF-N; line 6)

NDF Nitrogen represents nitrogen associated with the cell wall, measured by subjecting the Neutral Detergent Fibre (NDF) residue to the Kjeldahl procedure. NDF Nitrogen may be reported as:
- Neutral Detergent Insoluble Nitrogen (NDIN);
- Neutral Detergent Insoluble Protein (NDIP), or;
- Neutral Detergent Fibre Protein (NDF-P); expressed as a percentage of either total N, total Crude Protein or total Dry Matter.

A fraction of NDF-N will be both degradable by rumen microbes and digestible in the small intestine. A second fraction will be completely indigestible. It is generally assumed that the completely indigestible fraction is estimated as ADF Nitrogen x 6.25. The degradable/digestible fraction is, therefore, calculated by difference: [(NDF-N – ADF-N) x 6.25].

Starch (line 7)

Starch in feeds is analysed by first cleaving its long chains of glucose molecules with an enzyme (α-amylase) then oxidising the glucose with a second, indicator-linked enzyme.

Ash (line 7)

Ash is the residue remaining after complete combustion of the Crude Fibre + Ash fraction described in the Crude Fibre method described earlier. Ash represents the total mineral content of the feed.

Minerals (lines 8-13)

As stated earlier, most of the minerals found in plants are associated with organic compounds. Mineral analysis involves burning a sample of the feed, leaving an ash in which the minerals are present as simple inorganic (no carbon) salts. The ash is subsequently analyzed for each element specifically.

Mineral content may be determined by any of a number of alternative methods, including atomic absorption spectrometry, atomic emission spectrometry, mass spectrometry, neutron activation analysis, x-ray emission spectrometry, molecular light absorption spectrometry, molecular fluorometry, electrochemistry, combustion elemental analysis, volumetry, ion chromatography or gravimetry.

Notice on the feed analysis report that calcium, phosphorus, magnesium, potassium, sodium and chloride are given as percentage values with iron, manganese, zinc, copper and molybdenum are reported as mg/kg (milligrams per kilogram) and selenium as μg/kg (micrograms per kilogram). This is simply a reflection of the fact that some minerals are found in plants in relatively large amounts while others are present in only trace quantities. Some feed labs will report trace mineral levels in units of ppm or ppb:

1%	= 1 part per hundred
1 μg/g = 1 mg/kg	= 1 ppm (part per million)
1 μg/kg	= 1 ppb (part per billion)

Later, in the section on specific nutrient requirements, minerals will be divided into macromineral and trace or micromineral categories. This is not meant to imply that the minerals required in trace quantities are any less important than those required in larger quantities.

Mineral levels given on the feed analysis report cannot be considered in isolation, since many of the elements interact with one another, affecting availability (see p. 42). For example, a copper level of 8 µg/kg might appear adequate with reference to the requirement tables in Appendix A. However, if the molybdenum level is more than 2 µg/g, the availability of copper to the animal may be limited.

Vitamins

Vitamins are typically small organic molecules and, as such, they are usually measured by gas liquid chromatography (GLC) or high performance liquid chromatography (HPLC). Because these procedures are relatively expensive, vitamin levels in feed are not routinely measured.

Values calculated from analysis results

Non-Fibre Carbohydrates (NFC; line 14)

Non-Fibre Carbohydrates (NFC) represent feed carbohydrates, including starch, pectin and sugars, which are more rapidly degradable by digestive tract microbes relative to the cell wall carbohydrates measured as Neutral Detergent Fibre (NDF). NFC content is calculated by difference:

$$NFC\% = 100\% - CP\% - EE\% - NDF\% - Ash\%$$

A variation of NFC is Non-structural Carbohydrates (NSC; line 14), where:

$$NSC\% = 100\% - CP\% - EE\%$$
$$- (NDF\% - NDF\text{-}P\%) - Ash\%$$

Adjusted Crude Protein (ACP; line 15)

Adjusted Crude Protein (ACP) discounts CP to account for ADF-N. In most feeds, 3-8% of total CP will be associated with the ADF residue, even in the complete absence of heating. Therefore, most feed labs do not discount the total CP value for heat damage unless ADF-N values are excessive. Others assume that a fixed proportion (e.g., 70%) of ADF-N is unavailable.

Discounted CP (Total CP – ADF-P or Total CP – excess ADF-P) values are generally reported as Adjusted Crude Protein (ACP) or Available Protein.

Relative Feed Value (RFV; line 15)

Relative Feed Value is used as an index of forage digestibility and potential intake, calculated from ADF and NDF using the following equation:

$$RFV\% = \frac{[(88.9 - (0.78 \text{ x } ADF\%)) \text{ x } (120 / NDF\%)]}{1.29}$$

Digestible Protein (DP; line 16)

For sheep, Digestible Protein is calculated from CP assuming a true digestibility of 90 percent and a correction of 3 percent for metabolic fecal protein:

$$DP\% = (CP\% \text{ x } 0.9) - 3\%$$

Feed labs may use other calculations to estimate DP, based on recommendations for cattle. For example:

$$DP\% = 72.96\% - (1.02 \text{ x } ADF\text{-}P\% \text{ x } 100\%/CP\%)$$

Metabolizable Protein (MP; line 16)

For sheep, Metabolizable Protein is calculated as 70 percent of DP:

$$MP\% = (DP\% \text{ x } 0.7)$$

Protein Degradability (line 17)

Protein digestion in ruminant animals is a two-stage process (see figure 3.13, p. 34). First, a fraction of the dietary protein is broken down by microbes in the rumen to produce peptides, amino acids and ammonia. This fraction is referred to as either Rumen Degradable Protein (RDP) or Degradable Intake Protein (DIP). The dietary protein that escapes breakdown by rumen microbes (commonly called 'bypass' protein) is referred to as Undegradable Protein (UDP) or Undegradable Intake Protein (UIP).

Protein degradability is estimated by measuring the disappearance of crude protein from finely ground feed samples incubated in porous nylon bags in the rumen of a fistulated animal. CP remaining in bags removed from the rumen at fixed time intervals allows estimation of the rate of degradation.

The nylon bag procedure is laborious and expensive, limiting its use to research facilities. Results obtained with this method have been quite variable both within and between laboratories. Simpler benchtop methods have been proposed, in which feed samples are incubated with mixtures of protein-degrading enzymes extracted from the rumen. Although several commercial labs offer degradability analysis using these methods, lack of standardization of both protocols and enzymes makes it difficult to place confidence in results.

Total Digestible Nutrients (TDN: line 18)

The classical method of estimating TDN was based on the summation of digestible fractions determined by proximate analysis (p. 19):

$$\text{TDN\%} = \text{dig CP\%} + \text{dig CF\%} + \text{dig NFE\%} + (2.25 \times \text{dig EE\%})$$

The digestible ether extract is multiplied by 2.25 because, on oxidation, lipids provide 2.25 times more energy than the other fractions.

This method was very laborious and impractical for routine use, requiring the estimation of digestibility in animal feeding trials where each fraction was measured in feeds and feces over several days. Today, feed TDN value is calculated from ADF as described below for other energy fractions.

Gross, Digestible, Metabolizable, Net Energy (lines 18-20)

The relationships between various energy fractions are described in **Chapter 4** (pp. 38-39). Gross Energy (GE) potential of a feed is estimated by measuring the heat produced by complete combustion of a sample under controlled conditions in a 'bomb' calorimeter.

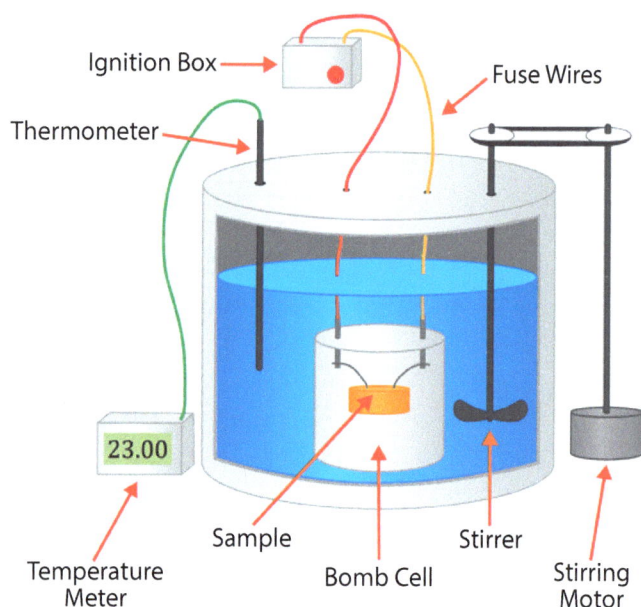

Figure 2.7: Schematic diagram of a bomb calorimeter.

Digestible Energy (DE) was originally estimated as the Gross Energy measured in the total amount of feed consumed over several days minus the Gross Energy measured in the total feces produced in the same time period, similar to the methods described for the determination of TDN.

Today, estimates of the digestible (DE), metabolizable (ME) and net (NE) energy values of forages are most commonly based on measurements of their Acid

Detergent Fibre (ADF) content. Since ADF represents the most indigestible fraction of a feed, it is assumed that the digestibility of a feed, and thus its energy-yielding potential, should be inversely proportional to its ADF content. Here are examples of regression equations used to calculate DE:

for legume forage:

DE, Mcal/kg = 3.91 – 0.036 x ADF%

for grass forage:

DE, Mcal/kg = 4.34 – 0.046 x ADF%

for mixed grass/legume forage:

DE, Mcal/kg = 4.08 – 0.040 x ADF%

Regression equations used vary from lab to lab, so it is important to determine whether differences in reported energy values between labs is due to different ADF values or different equations or both.

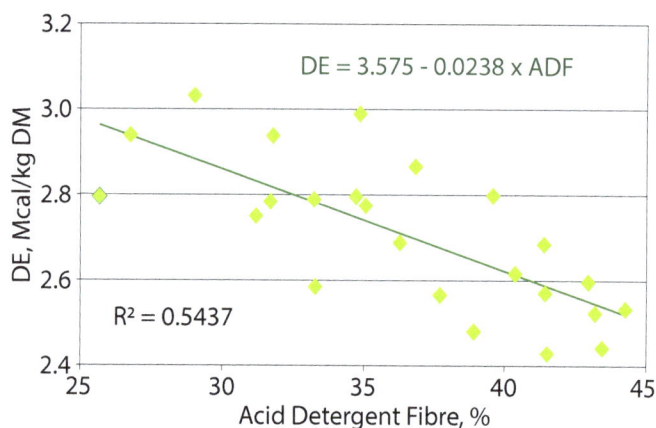

Figure 2.8: The relationship between Acid Detergent Fibre (ADF) and Digestible Energy (DE) for a set of Alberta grass forage samples. In this example, ADF value accounted for only 54% of the variation in DE values.

Estimates of the energy contributed by concentrate feeds are most commonly based on values published in tables of feed composition.

The following equation can be used to convert DE to ME:

$$\text{ME, Mcal/kg} = 0.818 \times \text{DE, Mcal/kg}$$

The conversion of ME to NE varies with the productive or reproductive state of the animal (maintenance, weight gain, gestation or lactation) and the energy density of the diet (see Appendix D).

Near infrared reflectance spectroscopy (NIRS)

NIRS is a quick, reliable, low cost, computerized method used to analyze feeds for their nutrient content. It uses infrared light instead of chemicals to identify important compounds and measure their amounts in a sample. Feeds can be analyzed in less than 15 minutes where chemical methods may take hours or days. This quick turnaround and the resultant cost savings in labour and consumables make NIRS an attractive method of analysis. The use of NIRS requires the establishment of calibration equations for each type of feed being analysed, relating specific reflectance wavelengths with chemical analysis results for the same feed type.

On-farm feed quality assessment

Dry Matter and Moisture

The laboratory assessment of feed dry matter and moisture content was described earlier (p. 19). When feeding ingredients such as silage or wet brewer's grains, it is often necessary to quickly check moisture levels to make sure the correct amount of ingredient dry matter is being included in the ration. The sidebar on the right describes how to check moisture levels with a microwave oven.

Silage pH

The proper preservation of silage is dependent upon its moisture content and acidity. The latter is measured on the scale of pH. Pure water has a pH of 7. A pH of 6 indicates a low level of acidity while pH 5 is ten times as acidic as pH 6 and pH 4 is again ten times as acidic as pH 5. pH values above 7 indicate alkalinity.

After fermentation is complete, high moisture silage (60–75% moisture; 25–40% DM) should have a pH below 4.5. The pH of haylage or low moisture silage may be slightly higher. Of particular interest to sheep producers, poorly preserved silage is often related to outbreaks of Listeriosis or circling disease. When high moisture forage is not well chopped prior to ensiling, it is impossible to eliminate sufficient air when packing.

Microwave oven dry matter estimation

Forage dry matter levels can be accurately estimated on-farm using an inexpensive microwave oven and an electronic postal scale. Mechanical postal scales are generally not accurate enough to measure gram differences in weights.

Here's how it's done :

1. Weigh a microwave-safe container large enough to hold 100-200 grams of wet forage (a paper bag is a good choice). Record the weight of the container (WC) or, if your scale has a tare adjustment, set the scale at zero (WC = 0).
2. Weigh 100-200 grams of wet forage into the container (WW). The larger the sample, the more accurate your determination can be.
3. Place a drinking glass or glass jar containing 250 ml of water in the back corner of the oven. The water serves as a 'ballast' to absorb excess energy, preventing ignition of the sample. If your sample does ignite, turn off the oven, unplug the power but don't open the door until the sample has burned completely.
4. Heat the forage sample at 80–90% of maximum power for 5 minutes. Re-weigh and record the weight.
5. Repeat step 4 until the weight is less than 5 grams lower than the previous weight.
6. Heat the sample at 30–40% of maximum power for 1 minute. Re-weigh and record the weight.
7. Repeat step 6 until the weight is less than 1 gram lower than the previous weight. This is the dry weight (WD).
8. Calculate Dry Matter (DM) % as follows:

$$\text{DM\%} = \frac{\text{WD - WC}}{\text{WW - WC}} \times 100$$

Improper fermentation results in a pH which is too high, providing good conditions for the proliferation of the bacteria which cause the disease.

Silage pH can be monitored with a simple, portable pH meter or with pH test strips.

Bulk density (bushel weight)

Bulk density is the weight of a standard volume of grain. In the past, this was measured as bushel weight but, since adoption of the metric system, kilograms per hectolitre (kg/hL) is used as the standard unit of measure. Table 2.1 provides bulk density standards for western Canadian feed grains. The Avery bushel used in Canada is equivalent to 36.37 litres. Be aware that, in the US, the 35.34 litre Winchester bushel is used.

Although bulk density is a generally accepted measure of grain quality, its relationship to nutritional value is not clear. Most assume that grain of higher density is worth more as a dietary ingredient due to the fact that the grain kernel is denser than the hull and also higher in digestible nutrients (mainly starch).

	lbs per bushel	kgs per hectolitre
Barley	48	62
Oats	37	48
Rye	49	63
Triticale	48	62
Wheat	51	65

Table 2.1: Bulk density standards for Canadian feed grains.

The practical implication of this relationship is very useful when a producer is buying grain without the benefit of a feed test. Light grain (e.g., barley weighing 42 lbs/bushel) should be worth less than heavy grain (e.g., barley weighing 50 lbs/bushel).

Factors affecting feed quality

Several factors that affect the feed test have already been mentioned:

- the energy content of feed grain is roughly proportional to bulk density (bushel weight);
- heating in storage reduces protein availability in hay and silage;
- improper preservation of silage results in lower energy and protein values.

In addition to these, stage of maturity, legume content and weathering can have major effects on forage quality.

The optimum time for harvest is a compromise between quality and quantity as shown in figure 2.9. The target will be based primarily on the nutrient requirements of the animals to which the forage will be fed. For example, during maintenance a 70 kg ewe will require a forage containing less than 9% crude protein, where the same ewe in early lactation nursing twins will benefit from forage with a much higher protein content.

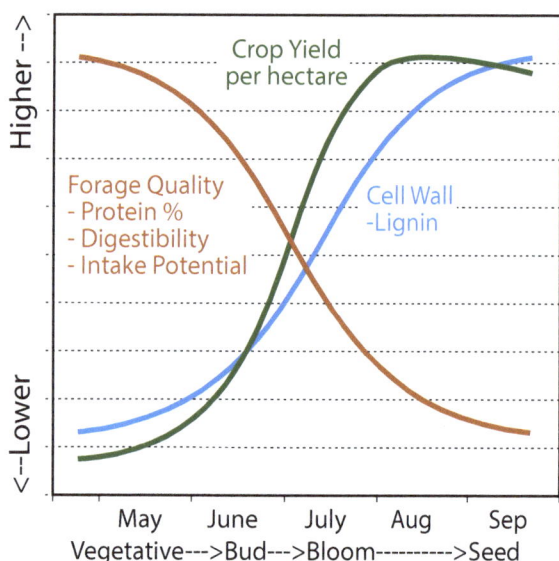

Figure 2.9: As grass or legume (this example) forage crops mature, quality declines as crop yield increases.

Stage of Maturity at Harvest	ME† Mcal/kg	% Crude Protein grass	% Crude Protein legume	Intake % BW
Vegetative	2.3	15	21	3.0
Boot or Bud	2.1	11	16	2.5
Bloom	1.8	7	11	2.0
Mature	1.6	4	7	1.5

Table 2.2: Effect of stage of maturity on the feeding value of grass and legume forages. †ME - metabolizable energy.

Table 2.2 illustrates differences in protein content of grasses and legumes at various stages of growth. Forages which have significant legume content have higher feeding values than those containing grasses alone. Finally, nutrients can be lost from hay during any stage of harvest or storage (table 2.3):

- cellular respiration: 4-15% of initial dry matter may be lost during wilting;
- legumes are particularly susceptible to leaf shattering if the crop has to be raked or tedded, and during baling, especially if the crop is very dry. Since leaves are significantly higher than stems in nutrient content at all stages of maturity (see figure 1.7), it is not surprising that leaf loss leads to lower forage quality.
- rain damage can leach nutrients from the crop as it lies in the field or in storage if it is not covered;
- vitamins, particularly vitamin A, may be lost due to UV exposure in the field;
- if moisture content exceeds 15–20%, heat produced by cellular respiration and/or microbial activity can result in the formation of heat-damaged protein (see ADF-N in **Feed Analysis** section, p. 22), mould growth and spontaneous combustion.

Harvest Stage	Legumes Dry Matter Losses (% DM)	Grasses Dry Matter Losses (% DM)
Mowing	1–3	1–2
Mowing and Conditioning	1–4	1–2
Tedding	2–8	1–3
Swath Inversion	1–3	1–3
Raking	1–20	1–20
Baling		
Small Rectangular	2–6	2–6
Large Rectangular	1–4	1–4
Large Round	3–9	3–9
Hay Storage		
Inside	3–9	3–9
Outside	6–30	5–22

Table 2.3: Ranges of dry matter loss at each stage of forage harvest. source: Alberta Agriculture.

Chapter 3: The Digestion of Feed

Earlier, it was pointed out that ruminants are relatively inefficient in the conversion of feed. The suggestion was made that the production of lamb (and beef) could only be economical when feed costs are low. Thus, economics dictate that sheep production be forage-based. The ability of sheep, and ruminants in general, to utilize forage is a function of their unique digestive system.

The digestive systems of mammals are broadly divided into three classes:

- Monogastric – meaning one stomach, includes: human, dog, cat, pig;
- Ruminant – includes: sheep, cattle, goats, deer, bison;
- Hindgut fermenters – includes: horse, rabbit.

In order to understand some of the unique properties of the ruminant system, it will be helpful to first briefly describe the simpler monogastric system.

Monogastric digestion

An outline of the pig's digestivesystem is shown in figure 3.1. Food, upon entering the mouth, is pulverized by chewing. At the same time, lubricating digestive juices containing enzymes are secreted from the salivary glands. These particular enzymes are responsible for initiating the breakdown of starch.

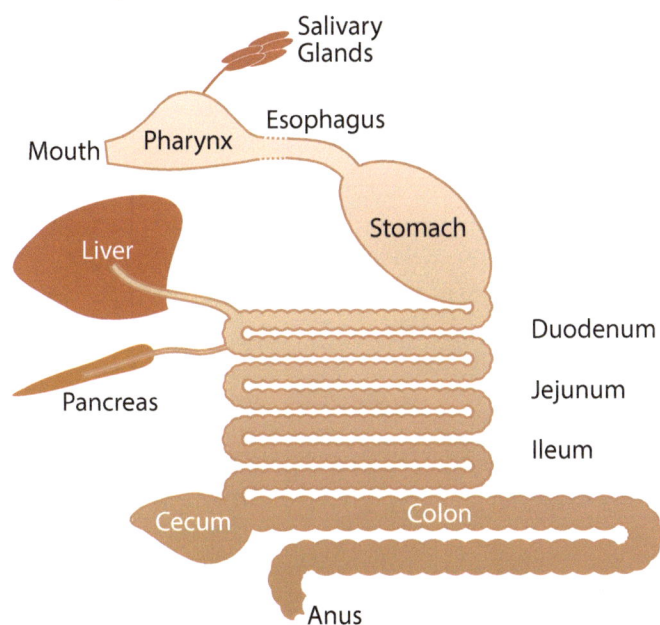

Figure 3.1: Schematic of the monogastric digestive system.

Food, now mixed with these secretions, passes down the esophagus into the stomach, where the digestion of protein, fats and oils is initiated by acid and other specific enzymes. In monogastrics, the stomach also serves as a reservoir for food which has been rapidly ingested.

Compartment	Digestive Volume (% of total tract)			
	Cattle	Sheep	Horse	Pig
Rumen	56.9	52.9		
Reticulum	2. 1	4.3		
Omasum	5.3	1.7		
Abomasum	6.5	7.7		
Total Stomach	70.8	66.6	8.6	29.2
Small Intestine	18.5	20.5	30.2	33.3
Cecum	2.8	2.6	15.9	5.6
Large Intestine	7.9	10.3	45.3	31.9
Total Capacity (litres)	356.0	44.0	211 .0	28.0

Table 3.1: Comparison of the relative sizes of digestive tract compartments in adult animals.

When the initial stages of digestion in the stomach are completed, the contents pass into the small intestine. Here, bile from the liver and gall bladder, as well as enzymes from the pancreas are added. Breakdown continues as the digesta travel the length of the small intestine, while at the same time the products of digestion are absorbed into the bloodstream.

Enzymatic breakdown of most of the organic constituents of food is complete by the time the unabsorbed digesta reach the cecum and large intestine. One of the main functions of this part of the system is the absorption of water and minerals of both dietary and secretory origin. Further breakdown of digesta is carried out here by a permanent population of microbes (bacteria, fungi and protozoa) with some proportion of the products being absorbed into the blood. Food material which has escaped both enzymatic and microbial digestion is excreted.

Hindgut fermenters, like the horse and rabbit, have a relatively large capacity for microbial digestion in the large intestine and a much-enlarged cecum. Table 3.1 compares the sizes of several parts of the monogastric digestive system. Notice that where the large intestine and cecum make up only 10.7% of the digestive system in cattle, they represent over 60% of the total volume in the horse. The implications of this will become clear when we discuss microbial digestion in the ruminant.

The sheep's digestive system

The main feature that distinguishes the monogastric digestive system from that of the sheep is the complex stomach illustrated schematically in figure 3.2. The first two sections, the reticulum and rumen, comprise a large fermentation compartment, referred to as the reticulorumen. These two sections are distiguishable by their uniqe linings—the inner surface of the reticulum

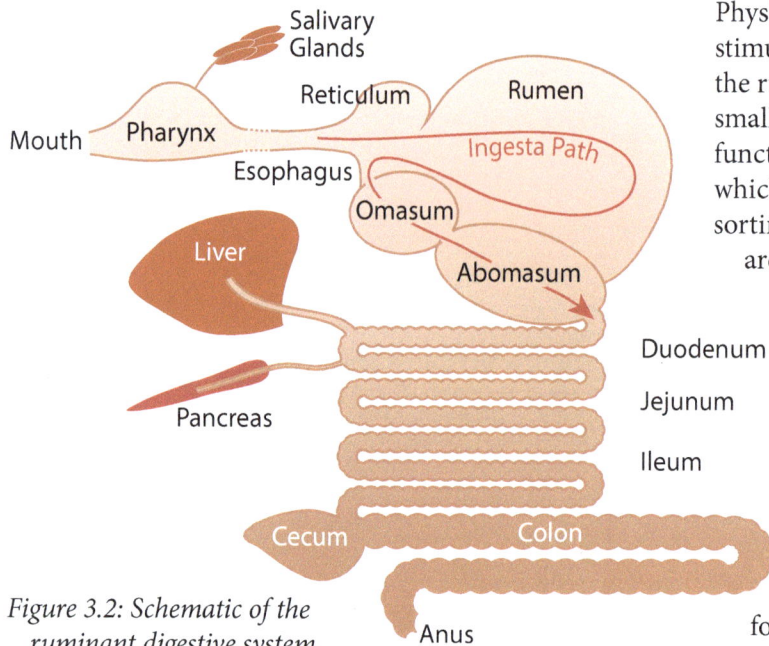

Figure 3.2: Schematic of the ruminant digestive system.

Physical fibre also provides a 'tickle factor' which stimulates rumen contractions. These help to keep the rumen contents well mixed and to force fluid and small particles further down the digestive tract. A third function of long forage is the maintenance of a fibre mat which floats in the rumen and functions as a particle sorting system. Long particles near the top of the mat are the first to be regurgitated for cud chewing which subdivides and adds water to the particles. When they re-enter the rumen, they 'float' at a lower level than the longer, drier particles they were derived from. A functional mat also stabilizes rumen fermentation by trapping fine particles, slowing their rate of breakdown by reducing exposure to rumen microbes.

The absence of adequate physical fibre in the ration to promote cud formation may account for the wool picking and wood chewing sometimes seen in feedlot lambs fed all-concentrate diets.

On entering the reticulorumen, feed is immediately subjected to microbial digestion. An extremely varied population of bacteria and protozoa (figure 3.4) attach themselves to the feed and begin the breakdown process. This is facilitated by the secretion of enzymes onto the feed and into the fluid contents of the rumen.

It should be noted that the microbial population that attaches to a particle of grain will be quite different from that which attaches to a forage leaf. This has important implications when changes in ration are contemplated. A slow transition is necessary to allow time for microbial populations to adapt to new feeds.

The convoluted inner surfaces of the reticulum and rumen serve two main functions. They vastly increase the area for absorption of nutrients and they also provide attachment sites for additional populations of bacteria. These bacteria, like the ones attached to feed

has a 'honeycomb' appearance whereas the lining of the rumen is like pile carpet having innumerable small, flat projections called papillae (see figure 3.3).

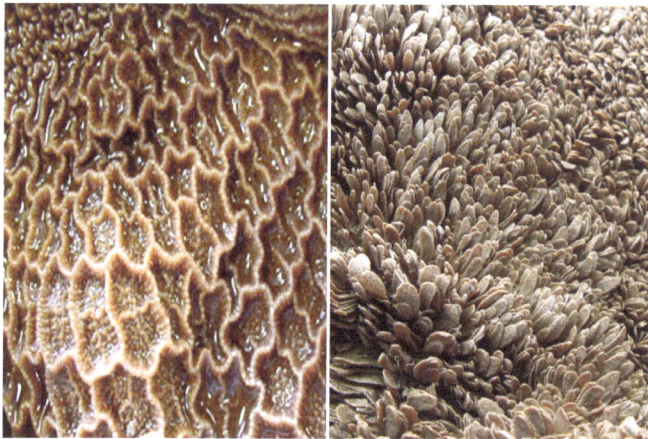

Figure 3.3: Inner surfaces (mucosa) of the reticulum (left) and rumen (right).

Ingested feed passes rapidly, with very little chewing, down a muscular esophagus into the reticulorumen. Later, boluses (cuds) of feed are regurgitated, broken down by chewing (rumination) and mixed with saliva.

Ruminant animals require 'physical fibre' in their diets to keep the rumen functioning normally. In addition to breaking down large forage fibre particles, chewing promotes salivation—the mature ewe produces up to 25 litres of saliva per day. This amount of fluid being produced high up in the digestive tract washes feed particles through the reticulum and rumen into the lower tract. In addition, saliva contains buffers which serve to prevent the contents of the rumen from becoming too acidic.

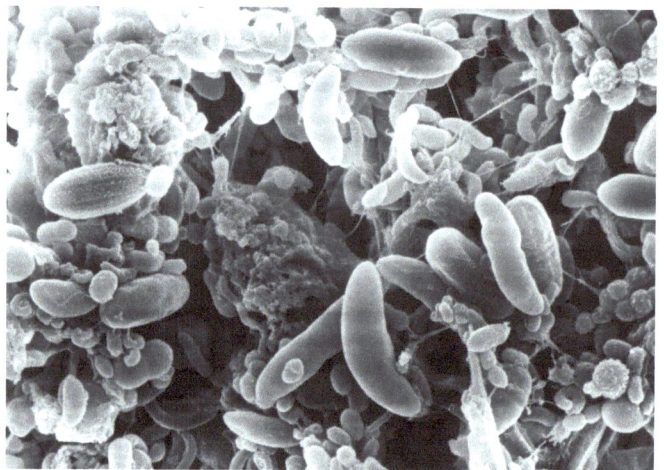

Figure 3.4: Scanning electron micrograph showing the variety of microorganisms found in the rumen.

particles, produce enzymes which are secreted into the fluid contents of the rumen. One of the important contributions of this particular population is the enzyme urease which is responsible for the breakdown of urea. So feed is subjected to digestion both by enzymes dissolved in the general milieu of the rumen and, more specifically, by those produced by attached microbes.

Continual mixing of rumen contents is essential to efficient fermentation. The muscular walls of the rumen and reticulum produce waves of contraction traveling their combined lengths at about half-minute intervals. This process, in addition to mixing the rumen contents, facilitates both regurgitation for further 'cud-chewing' and belching, which releases gases produced by fermentation (mainly hydrogen and methane). Under some conditions (e.g., grain overload) the muscular walls may stop contracting resulting in rumen stasis, which can place the animal at serious risk of bloating.

After the feed has been sufficiently chewed and broken down by microbial action, the digesta enters the omasum. Flow into this third segment of the ruminant stomach is regulated by a small opening called the reticulo-omasal orifice which prevents large particles from leaving the rumen. It is the small caliber of this orifice which makes it possible for sheep to utilize whole grains. The larger orifice in cattle allows particles the size of whole grain to pass into the lower gut and be excreted.

The omasum itself is a muscular organ having multiple internal leaves (laminae) like the pages in a book (see figure 3.4), providing a large internal surface area. It is thought to have two main functions. The first is the absorption of water, electrolytes and products of microbial digestion from the rumen fluid, yielding a product for further digestion which has a significantly higher proportion of dry matter.

Figure 3.4: The 'many-leaved' lining of the omasum (left) and the mucosal lining of the abomasum (right).

Secondly, the omasum serves as a pump, propelling digesta from the reticulorumen into the fourth segment of the stomach, the abomasum.

The ruminant abomasum is analogous to the true stomach of the monogastric with its digestive processes being very similar to those described earlier for the pig. Digestion and absorption of its products progress as the digesta passes down the small intestine.

The large intestine and cecum of the sheep represent only about 12% of the total volume of its digestive system. This may seem relatively insignificant in comparison with the horse (table 1.1). However, fermentation in this area can make a significant contribution to overall digestion. This will be discussed further in the section on **Carbohydrate digestion** (p. 31).

Development of the ruminant stomach

At birth, the lamb's rumen and reticulum have a capacity roughly equal to that of the abomasum (figure 3.5). They contain no micro-organisms and, as a consequence, are not capable of functioning as they do in the adult. Bacteria begin to populate the rumen shortly after birth as the lamb begins to nurse and explore its environment. However, it takes several weeks before a stable microbial population is established which is capable of efficient digestion.

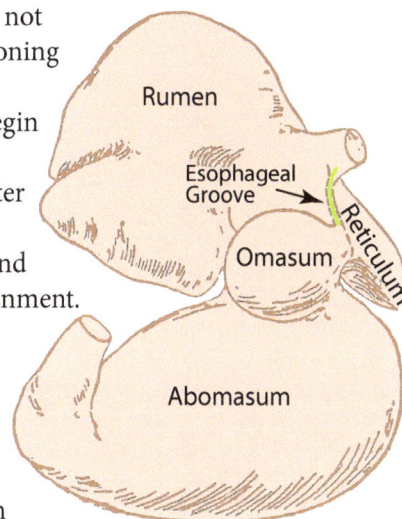

Figure 3.5: The stomach of the newborn lamb.

Attempts have been made to hasten the establishment of functional microbial populations in newborn lambs by administering mixed microbes from the rumens of older animals. There are two reasons for this. First, many of the bacteria which contaminate the digestive tract from the environment early in life are capable of producing digestive upsets. By inoculating the rumen and reticulum with a more appropriate microbial population, competition might protect the digestive tract from the adverse effects of such contaminants.

The second reason for attempting to establish a functional population is to hasten the ability of the rumen and reticulum to digest solid feed. This would make it possible to wean lambs earlier, a particular advantage when accelerated lambing is being attempted.

The esophageal groove

Since the rumen and reticulum are non-functional in the newborn lamb, a mechanism has evolved which allows milk to flow directly to the omasum. A reflex reaction causes a muscular fold on the wall of the reticulum to form a closed tube leading from the end of the esophagus to the reticulo-omasal orifice (see figure 3.6). This fold is called the esophageal groove and an appreciation of its function will affect some of the management aspects of feeding newborn lambs.

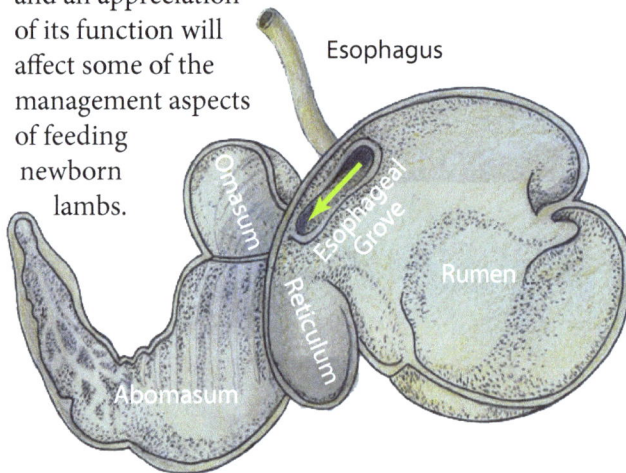

Figure 3.6: The esophageal groove channels milk directly from the esophagus into the omasum.

The esophageal groove closes in response to behavioural stimuli associated with the ingestion of liquid feed such as nursing from the ewe or feeding from a nipple pail. Even the sight of a nipple bottle may elicit the response in an orphan lamb. However, the reflex requires some degree of training. Therefore, it is used to best advantage when feeding routines are well established.

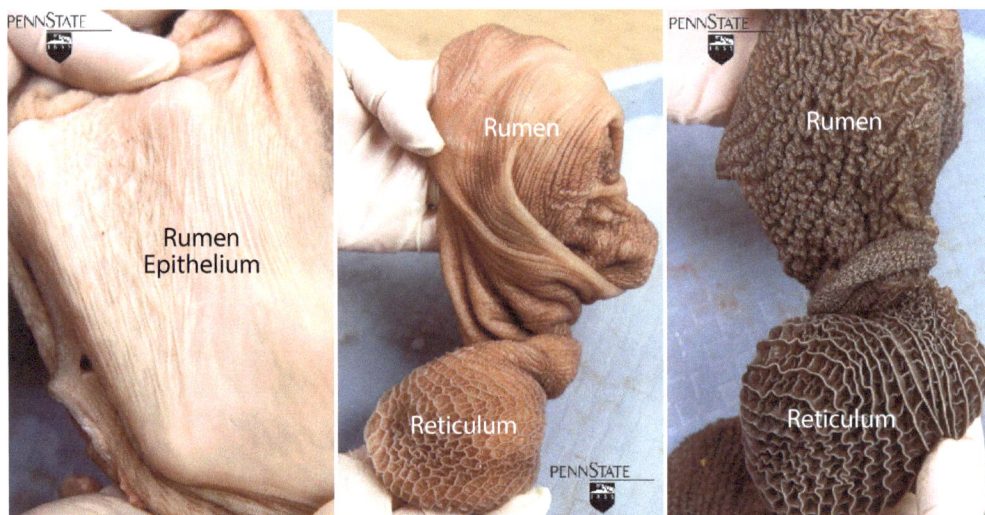

Fgure 3.7: Differences in development of the rumen in 6 week old calves fed milk only (left), milk and hay (centre) or milk and grain (right). source: Penn State University

If at all possible, weak newborn lambs should be encouraged to suckle either from the ewe or from a bottle. Although feeding by stomach tube may be the only alternative in some cases, this will invariably result in milk spilling into the reticorumen. A similar situation arises when milk is ingested too rapidly to be accommodated by the esophageal groove. This can occur when milk replacer is fed from a bottle or from the bottom of a nipple pail where a round-holed (rather than a crosscut) nipple is used.

Milk that finds its way into the rumen and reticulum is subjected to fermentation by bacterial contaminants early in life. Such fermentation may result in significant gas production resulting in a typical pot-bellied lamb. The young lamb cannot expel this gas efficiently since the belching mechanism is poorly developed.

Lambs being fed through a rubber nipple should be encouraged to suck. Frequent feedings of small volumes are usually more successful than large volumes fed infrequently. These management considerations are discussed in more detail later (pp. 55–56).

Effect of feeding management

Between birth and maturity, the rumen and reticulum increase tenfold in volume in relation to the abomasum; the rate at which this proceeds can be significantly altered by nutritional management.

Most newborn lambs show little interest in consuming solid feed before they are two or three weeks of age. Consequently, until that time they must be nourished exclusively with milk or milk replacer. After this time it is possible to accelerate rumen development through feeding practices.

The closure of the esophageal groove only occurs when liquid feed is ingested. Therefore when solid feed is consumed it travels directly to the rumen where it is fermented to produce volatile fatty acids (VFA; see section on **Carbohydrate digestion**, p. 31). In the past, it was commonly believed that feeding hay to young lambs would promote rumen development. The rationale was that physical 'scratch' was needed to start the rumen working. It is now known that the main stimulus to rumen development is VFA production from feed fermentation. Because the amounts of VFA produced from grain are higher than those from forage, the rumen develops much faster when grain-based diets are fed, as shown in figure 3.7. For this reason, hay should not be offered to lambs until after they are weaned, when fibre is required to promote the growth of the muscular layer of the rumen and to maintain the health of the papillae. Rumen papillae can grow too rapidly in response to high levels of VFA. When this happens, they may clump together, reducing the surface area available for absorption. Also, some 'scratch' is needed to keep the papillae from forming layers of keratin (skin-like tissue), which can also inhibit VFA absorption.

Creep feeding has become common practice in most successful sheep operations. The aim is to provide palatable, high quality solid feed to encourage consumption as early in life as possible. Restriction of milk intake after solid feed consumption is well established further promotes the intake of creep ration. The higher the quality and the greater the quantity consumed, the higher will be the rate of VFA production and the more rapid will be the rate of rumen development. In addition to maximizing growth potential by increasing the lamb's ability to consume nutrients, creep feeding also facilitates early weaning in accelerated lambing systems.

Carbohydrate digestion

As pointed out earlier, forages contain only low levels of fats and oils. Consequently, the main sources of energy for the ruminant are the carbohydrates.

Monogastric carbohydrate digestion

Carbohydrate digestion in the monogastric begins when food is mixed with saliva containing enzymes (amylases) which begin the breakdown of starch. The process continues in the small intestine, facilitated by pancreatic amylases. As digestion progresses, the end products (the simple sugars: glucose, galactose, etc.) are absorbed into the bloodstream. Depending on the energy status of the animal, the sugars may be used as immediate energy sources or stored as glycogen for later use. In the lactating female, glucose is primarily used in the manufacture of the milk sugar, lactose.

Ruminant carbohydrate digestion

The enzymes which mediate carbohydrate breakdown in the ruminant are mainly of microbial origin. Each of the several classes of carbohydrates is digested by specific enzymes produced by a distinct microbial population (figure 3.8).

Volatile fatty acids

If oxygen were present in the rumen, the end-products of carbohydrate digestion would be carbon dioxide and water, the compounds from which the carbohydrates were originally synthesized in the plant (p. 3). However, the microbial population in the rumen operates in the absence of oxygen (the rumen environment is *anaerobic*), resulting in incomplete carbohydrate breakdown. Under these conditions, the principal end-products of digestion are compounds referred to as the volatile fatty acids (VFAs), including acetic acid, propionic acid and butyric acid. Incomplete breakdown in this case is analogous to the situation where wood is burned with limited air. The smoke produced is composed of the products of incomplete combustion.

The breakdown of carbohydrates to VFAs results in the release of significant amounts of feed energy. This energy is utilized in the rumen for microbial growth involving, for example, the synthesis of new microbial protein, fat and carbohydrate. The VFAs are absorbed into the bloodstream through the wall of the rumen to serve as energy sources for the animal itself.

Other important products of ruminant carbohydrate digestion are the keto acids. These are formed in much smaller quantities than the VFAs, but serve an important role in microbial protein synthesis (pp. 33–34). Methane, also a major end-product of anaerobic fermentation, is largely lost to the atmosphere.

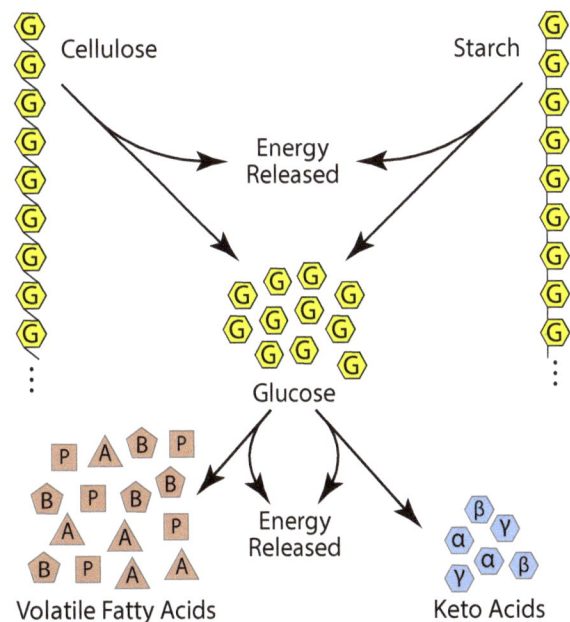

Figure 3.8: In the ruminant, microbes initially break down both fibrous and non-fibrous carbohydrates to produce simple sugars which are subsequently fermented to produce VFAs, keto acids and other end-products.

Carbohydrates that escape digestion in the rumen and those associated with the microbial population may be utilized further down the digestive tract. For example, carbohydrates that are components of microbial cells are digested and the resulting sugars absorbed in the small intestine, as in monogastrics. The contributions of these simple sugars to the overall energy requirements of the sheep are minor.

Further carbohydrate breakdown may also occur in the large intestine. Experimental results suggest that 5–10% of the energy requirement of lambs could be met from VFAs produced by the microbial population found here.

Cellulose digestion

The microbial populations of both the rumen and the cecum produce the enzyme cellulase which is responsible for the breakdown of cellulose. It is this feature which makes ruminants and hindgut fermenters like the horse unique in their ability to utilize forages in the production of meat, milk and fibre. Mammals are incapable of digesting cellulose without the aid of these microbes. Two features of cellulose utilization should be appreciated when feeding sheep:

The digestion of cellulose is a relatively slow process. One implication of this is the fact that feed consumption is limited by the rate at which feed is digested. Feed can be introduced into the rumen only as rapidly as the products of digestion are discharged. This means that rations high in fibre will be consumed in lower quantities than those that are low in fibre.

A second consideration is the degree to which the cellulose in a feed is associated with lignin. As forages mature, the cellulose found in the plant's cell walls becomes more lignified. The result is that the cellulose becomes less digestible. This is reflected in lower energy values and intake potential for forages as they mature (table 2.2).

Other carbohydrates

Starch, pentosans and simple sugars are rapidly fermented in the rumen to produce VFAs, keto and lactic acids, CO_2, hydrogen and methane. Under most conditions, lactic acid production is low. However, when feeds containing large quantities of readily fermented carbohydrates are rapidly consumed, lactic acid production may be significant, resulting in ruminal acidosis (pH below 5.5). Since stable pH is critical to proper rumen function, rapid changes can cause digestive problems. For example, grain overload results in depressed rumen pH which can lead to rumen stasis. The animal becomes unable to expel fermentation gases by belching and bloat occurs.

One of the main advantages of feeding whole grain to sheep lies in the reduced rate of acid production since the starch is not as immediately available to microbial breakdown as it is in processed grain. This results in more prolonged digestion and a more stable rumen pH. The effects are also seen in the health of the rumen papillae which can become damaged (figure 3.9) with long-term exposure to low pH.

Figure 3.9: Healthy (left) and acidosis-damaged (right) rumen papillae.

Protein digestion

Proteins, as described earlier, contain carbon, hydrogen, oxygen, nitrogen, usually sulfur and sometimes phosphorus. In order to understand protein digestion, a brief description of their chemistry is required.

Amino acids

Each different protein consists of a unique combination of 20 amino acids. In turn, each amino acid contains a single atom of nitrogen in combination with two atoms of hydrogen called an amino group (NH_2). Not all proteins contain all amino acids. These concepts are illustrated in figure 3.10.

Monogastric protein digestion

When protein is digested by the monogastric animal (figure 3.11), the long chains are first broken down into shorter chains called peptides, a process which begins in the stomach. In the small intestine, further digestion releases the individual amino acids which are absorbed into the bloodstream. The animal now uses these absorbed amino acids as building blocks for its own particular types of protein.

Protein quality

Because the animal has specific requirements for individual amino acids, the concept of protein quality arises. If the balance of amino acids in the feed protein is very similar to that required by the animal, the protein is said to be of high quality (table 3.2).

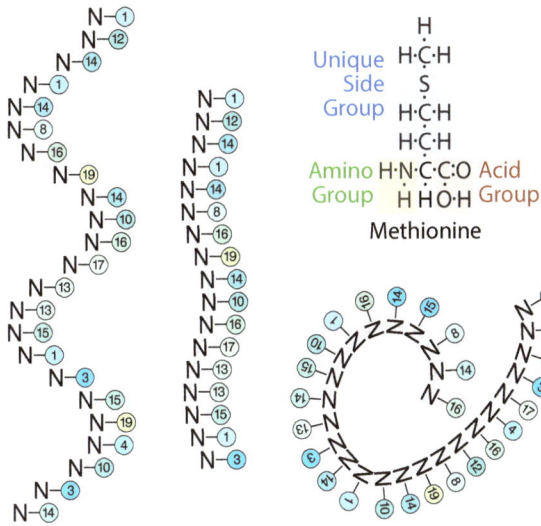

Figure 3.10: There are 20 different amino acids, each characterized by its unique side group (methionine illustrated here). Proteins are composed of long chains of up to thousands of amino acids, each different protein characterized by its unique amino acid sequence and shape.

If the amino acid composition of feed proteins is poorly matched to requirements, the protein is of low quality. Surplus amino acids are broken down in the liver and kidney. The amino group is removed and may be recycled or excreted in the form of urea while the remainder of the molecule is used as an energy source.

Ten of the 20 amino acids can be synthesized by mammalian tissues as long as a source of nitrogen is available. For this purpose, nitrogen may be obtained from urea or from the amino groups of surplus amino acids. The remaining 10 *essential* amino acids must

be supplied in the diet of the monogastric animal. It is the limited availability of these essential amino acids which usually restricts the synthesis of animal proteins. For example, when corn is fed to pigs, the limited availability of the essential amino acid lysine usually restricts growth. When corn-based rations are supplemented with lysine, a significant improvement in feed conversion efficiency occurs (see **Balanced rations**, p.48).

Protein Source	Protein Quality
Grass Forage	38
Legume Forage	52
Corn Grain	58
Wheat Grain	59
Canola Meal	62
Mixed Microbes	65
Soyameal	70
Oats Grain	70
Barley Grain	72
Fishmeal	77
Milk	87
Egg	99

Table 3.2: Protein quality expressed as relative biological value.

Ruminant protein digestion

Protein digestion in the ruminant is much more complex than that in the monogastric. Plant proteins enter the rumen and reticulum where they are attacked by the microbial population and broken down to their constituent amino acids. However, microbial degradation does not stop here. Most of these amino acids are further degraded, resulting in the release of the amino group which, when combined with a third hydrogen atom, forms ammonia (NH_3) as shown in figure 3.12. As was the case in carbohydrate digestion, protein degradation releases energy.

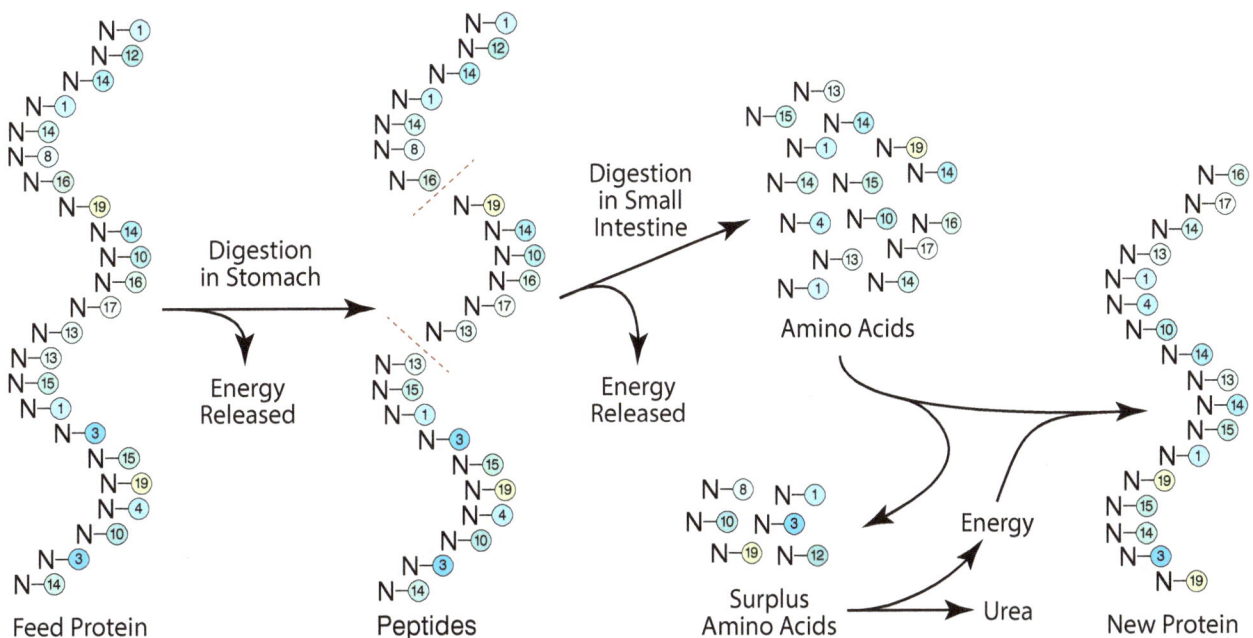

Figure 3.11: Protein digestion in the monogastric animal.

Ammonia in the rumen can also be derived from non-protein nitrogen (NPN) sources. Mention was made earlier of the large quantities of saliva produced by the sheep. This saliva serves as a vehicle for recycling amino groups in the form of urea (figure 3.13) back into the digestive system from amino acid breakdown elsewhere (e.g., liver). Urea may also be included in the feed as an inexpensive source of crude protein. Urea is broken down in the rumen by the enzyme urease and its amino groups are released as ammonia. Reference has already been made to the fact that the bacterial population adherent to the rumen papillae is a major contributor of urease (p. 29). The practical implications of this will be discussed later.

The first stage of nitrogen digestion in the sheep is completed with the production of amino acids and ammonia, accompanied by the release of energy. The second stage involves the use of these products by the bacteria, protozoa and fungi in the rumen to build new microbial protein. This process, illustrated in figures 3.12 and 3.13, requires recapturing some of the energy released in the first stage as well as energy released

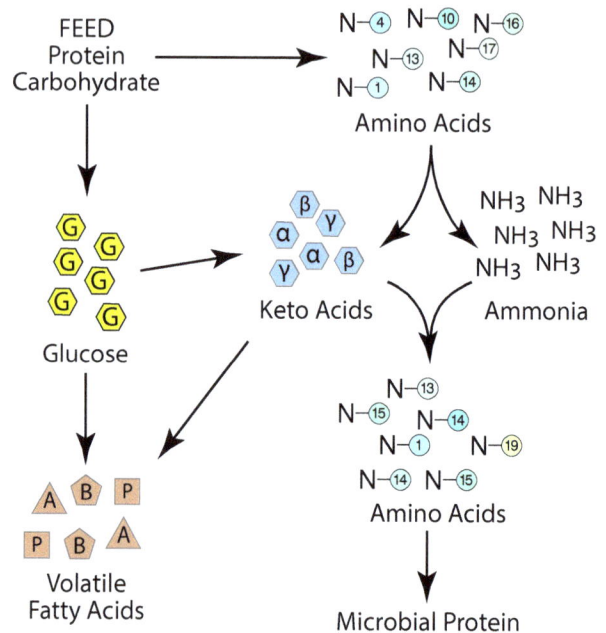

Figure 3.12: Protein digestion in the ruminant.

from carbohydrate breakdown. The keto acids released during both protein and carbohydrate digestion are combined with ammonia to form new amino acids and, subsequently, new microbial protein.

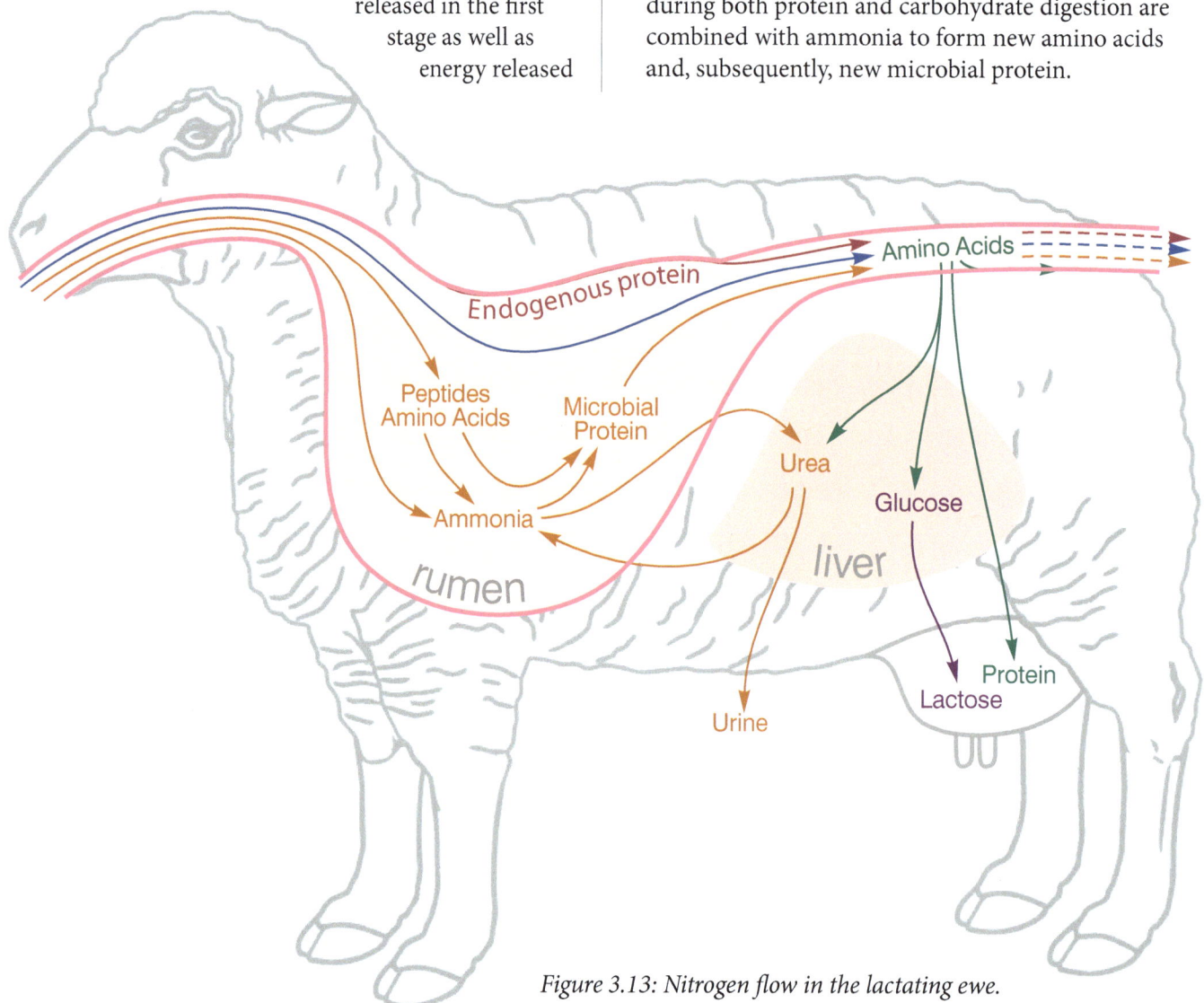

Figure 3.13: Nitrogen flow in the lactating ewe.

Having completed the processes of plant protein degradation and microbial protein synthesis, micro-organisms are drawn through the omasum and into the abomasum. Here the digestion of microbial protein begins in a manner similar to that in the monogastric. Digestion continues in the small intestine with amino acids being absorbed into the bloodstream to provide building blocks for animal protein.

Since the microbial population in the rumen has the capacity to synthesize essential amino acids, proteins that are of lower quality for monogastrics are actually improved by the rumen microbes. On the other hand, higher quality proteins are also modified with the result that the protein reaching the small intestine is of a relatively uniform medium quality irrespective of the protein in the feed unless a significant proportion of undegradable intake protein has been provided (p. 23).

As described above, protein digestion in the ruminant involves two extra steps compared with that in the monogastric. Each extra step in the digestive process results in some loss of overall efficiency. This, combined with the fact that microbial protein is of mediocre quality, means that sheep are inefficient in utilizing high quality protein in comparison to monogastrics. Since high quality protein sources are usually expensive, these observations reinforce the concept that sheep rations must be based on inexpensive ingredients, primarily forages.

Urea in sheep rations

As mentioned earlier, urea appears in the rumen in association with saliva. Urea and other non-protein nitrogen (NPN) sources may also be added to sheep rations to increase the level of crude protein. However, to make efficient use of urea it is necessary to appreciate a few features of its metabolism.

Urea and other sources of NPN added to sheep rations are not efficiently utilized until the rumen microbial population becomes adapted to their increased availability. Although this process begins within a few days of the introduction of additional NPN, it may take several weeks before maximum utilization is attained. Therefore, short-term feeding of NPN supplements makes little sense. In addition, for best results, NPN should be used to contribute no more than one-third of the total crude protein content of the complete diet.

The micro-organisms that produce urease (the enzyme responsible for urea breakdown) are concentrated on the inner lining of the rumen. Consequently, when large quantities of urea are fed over short periods of time, high concentrations of ammonia can accumulate near the rumen walls. This ammonia can produce a rapid

increase in rumen pH and, after passing into the blood stream, can cause alkalosis (high blood pH). The effect is opposite to the effect produced by grain overload.

In order to be incorporated into microbial protein, the ammonia produced by urea degradation must be combined with organic keto acids to form amino acids (figure 3.12). In addition, protein synthesis requires far more energy than is released by urea breakdown. It is important, therefore, when feeding urea to also provide readily fermentable carbohydrates as a source of both keto acids and energy. In fact, this principle applies to the efficient utilization of ammonia in the rumen irrespective of its source. Rations should contain approximately 4.4 Mcal of digestible energy (DE) for every 120 grams of crude protein degraded in the rumen (degradable intake protein; DIP).

When ammonia is produced in excess of the availability of keto acids and energy, it is absorbed through the walls of the rumen into the bloodstream. To some extent it may be recycled into saliva in the form of urea, but the greater proportion will be simply excreted (figure 3.13).

The emphasis above was on readily fermentable carbohydrate. It is important that the rate of carbohydrate degradation be well matched to the rate of urea degradation when animals are fed on a periodic basis (i.e., not self-fed). Cellulose, for example, is inappropriate for this purpose because of its slow rate of digestion whereas the starch found in feed grains is ideal (figure 3.14). Self feeding, where feed is consumed in frequent, smaller meals results in more stable conditions in the rumen and these considerations become somewhat less important.

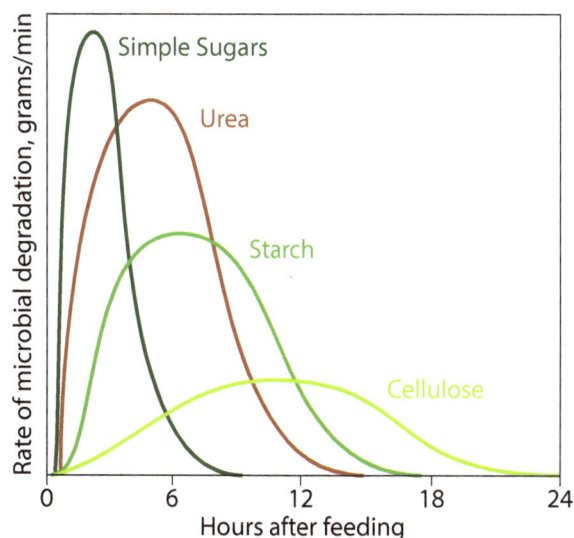

Figure 3.14: Rumen microbial degradation rate of urea compared with rates for various carbohydrates.

Undegradable intake protein

It was suggested earlier that most of the feed protein entering the rumen is degraded to ammonia. In fact, the degree of degradation varies, depending on the source of protein (table 3.3). For most of the grass and legume forages, protein degradability (degradable intake protein as % of crude protein) is in the 65 to 82% range; degradability of oats and barley protein is in a similar range. Among the crude protein supplements, urea is considered 100% degradable; canola meal protein is 57 to 68% degradable while soymeal protein degradabilty is slighly higher. Fish meal protein, which is rarely used in sheep rations, is the extreme at approximately 34 to 50% degradability.

Protein that is not degraded in the rumen is termed undegradable intake protein (UIP). Under some circumstances it is possible to use UIP to improve the overall quality of protein reaching the small intestine. Amino acids released from the UIP upon digestion there may complement the amino acids released from microbial protein, resulting in a better balance of amino acids being absorbed into the bloodstream. This results in more efficient overall utilization of feed protein.

The concept of protein degradability explains why some feeds yield better performance results than others even though the total amount of crude protein provided in the ration is the same. For example, production is often improved when soymeal or fish meal is used in place of urea. In a summary of trials where various sources of UIP were included in rations for lactating dairy cows, fish meal consistently produced higher milk yields. This was likely due to the UIP in fish meal providing a better complement of amino acids to those provided by microbial protein.

Protein Source	DIP, % of CP
Grass Hay	65–77
Grass Silage	70–82
LegumeHay	78–81
Barley Straw	53–68
Oats Grain	80–85
Barley Grain	67–76
Corn Grain	48–53
Soymeal	65–70
Fishmeal	34–50
Corn Distillers Grains	45–55
Corn Gluten Meal	30–40
Canola Meal	57–68
Urea	100

Table 3.3: Degradable intake protein (DIP) as % of crude protein (CP) in some common sheep feeds.

Heat-damaged protein

When silage is poorly packed, excess air can result in significant heating. Similarly, hay which is baled when its moisture content is too high may produce heat. Heating is caused by chemical reactions in the feed which result in part of the protein combining with carbohydrates. The compounds formed are less digestible by rumen microbes. When heat damage is slight, these compounds may be digestible in the abomasum and small intestine serving a role similar to that of UIP. However, in many cases, heat damaged protein will be totally indigestible, with the amount of loss being proportional to the degree of heating. A chemical test (acid detergent fibre nitrogen) to determine the degree of loss was discussed in the section on Feed Analysis (p. 22).

Chapter 4: Nutrient Requirements

In North America, nutrient allocations for livestock are based on the recommendations of the US National Research Council (NRC). The most recent NRC recommendations for sheep are published in the 2007 publication, 'Nutrient Requirements of Small Ruminants' which also includes recommendations for goats, deer, llamas and alpacas. The focus of that publication is on grazing which is the most common production environment under which these animals are raised worldwide, in contrast to the confinement and semi-confinement environments found in western Canada.

The previous NRC publication that focused exclusively on sheep was the 1985 'Nutrient Requirements of Sheep'. Based on the author's judgement, the dietary energy and protein recommendations stated in the 1985 publication are more appropriate for the available feeds and production systems employed in this part of the world. Therefore, the recommendations given in the present publication are based on a combination of those given in the 1985 and 2007 NRC publications, supplemented with information from the other sources listed in Appendix E.

Before discussing specific nutrient requirements, it will be useful to classify and describe the essential nutrients and, at the same time, point out the interrelationships between nutrient classes.

Essential nutrients

Water

This is an essential nutrient which is often overlooked in designing feeding programs for livestock. Its importance is illustrated by the following anecdote:

In 1978, the Alberta Ram Test Station was housed in a closed building on the Calgary Stampede grounds. Although it was well ventilated, it tended to be a little warm. A complete pelleted ration was designed which would promote maximum growth of the lambs on test. Water was provided in automatic, float-operated fountains. The first groups of lambs got off to a rapid start and grew very well for about six weeks (they were weighed every two weeks). After that, growth began to slow and feed consumption decreased, to the dismay of the management, who began looking for causes.

Initially the feed was blamed and a new lot was brought in. Next, the temperature in the barn was considered. Finally, the water was examined. Although the water in the drinking bowl had been kept clean, no attempt had been made to clean the float compartment. Here a healthy population of micro-organisms had developed which was imparting an off-flavor to the water in the drinking bowl. Consequently the lambs were limiting their water intake which was affecting their appetite and thus, their growth. As soon as the problem was rectified, rapid gains returned.

The point is this: sheep are fussy drinkers. In addition, if they are forced to subsist on an inadequate water supply in periods when their nutrient requirements are high, production may suffer. As in the case above, insufficient water intake may result in reduced dry matter consumption and often affects salt and mineral intake. In late pregnant ewes, whose requirements are particularly high, inadequate water intake leading to reduced feed consumption may result in pregnancy toxemia.

As a demonstration of the benefit of a clean, fresh water supply, feed consumption of feedlot lambs at Fairview College in Alberta was increased by providing water which was constantly circulated past the lambs thorough a rain gutter. The increased intake was apparently the result of appetite stimulation through increased water consumption.

Waterers which are not kept clean are also a source of disease. Several common health problems which result from fecal contamination are spread through fouled water supplies. Small-volume waterers that provide fresh water on demand are preferred over large water troughs that accumulate contaminants and need to be completely emptied to be effectively cleaned.

Although water may appear clean, various dissolved minerals and agricultural chemicals can also limit its palatability and consumption. Table 4.1 provides guidelines for maximum tolerable mineral levels in livestock water sources.

Several other factors affect the water consumption of sheep, including age, stage of production, wool cover, water temperature and environmental temperature and humidity. Table 4.2 demonstrates the influence of ambient temperature on estimate water intake for various classes of sheep.

Figure 4.1: The ideal sheep waterer will automatically maintain a small volume of fresh water, prevent freezing and be easy to clean.

Parameter	Max ppm	Parameter	Max ppm
Major Ions and Nutrients			
Calcium	1000	Sulphate	700
Nitrate plus nitrite	100	Total dissolved solids	3000
Nitrite alone	10.0		
Heavy Metals & Trace Ions			
Aluminum	5.0	Iron	NGE
Arsenic	0.025	Lead	0.1
Boron	5.0	Manganese	NGE
Cadmium	0.08	Mercury	0.003
Chromium	0.05	Molybdenum	0.5
Cobalt	1.0	Nickel	1.0
Copper	0.5	Selenium	0.05
Fluoride	2.0	Zinc	50.0

Table 4.1: Maximum acceptable concentrations of solutes in drinking water for sheep. NGE: No guideline established, ppm: parts per million = mg/litre.

The consumption of clean snow can account for a significant proportion of water intake in winter, but should not be relied upon as the only source. Sheep consuming feeds that are high in water content (silage, pasture) will satisfy a sizable proportion of their water requirements from their feed.

	Ambient Temperature		
	below 16°C	16–20°C	above 20°C
Class of Sheep	Water intake (litres/kg DM consumed)		
Lambs (up to 4 weeks)	4.0	5.0	6.0
Sheep (growing or adult, non-pregnant, dry)	2.0	2.5	3.0
Pregnant Ewes			
mid pregnancy			
single	2.5	3.1	3.7
twins	2.8	3.5	4.2
late pregnancy			
single	3.3	4.1	4.9
twins	4.4	5.5	6.6
Lactating Ewes			
first month	4.0	5.0	6.0
second month	3.0	3.7	4.5
late lactation	2.5	3.1	3.7

Table 4.2: Estimated water intakes for sheep.

Energy

Inadequate energy limits performance more often than any other nutrient deficiency. In lambs, the symptoms are most commonly slow growth and lower resistance to infection. Ewes may experience loss of weight, reduced fertility and lamb birth weights, inadequate milk production, shortened lactation periods and reduced wool quantity and quality. Energy deficiency may be a result of insufficient feed intake or, more commonly, low dietary energy concentration.

As suggested earlier, carbohydrates, fats, oils and proteins can all serve as energy sources although the most significant contribution for ruminants is from carbohydrates.

Gross Energy (GE)

As described in **Chapter 2**, the GE content of a feed is estimated by burning a sample in a 'bomb' calorimeter and measuring the amount of heat produced.

Digestible Energy (DE)

Of the total energy potentially available from the feed consumed, some is released in the digestive process or contained in metabolites that are absorbed into the bloodstream. The remainder is contained in indigestible compounds like lignin which are excreted (figure 4.2). In the feed analysis process, the digestible portion of feed energy is calculated as DE from the acid detergent fibre (ADF) value (p. 24).

Some of the energy which is apparently digested is actually lost as gas from the rumen when the animal belches. A second portion is lost through urinary excretion.

Metabolizable Energy (ME)

Digestible energy minus gaseous and urinary losses is termed metabolizable energy (ME). For most feeds, the assumption is made that ME amounts to 82% of DE.

Net Energy (NE)

Metabolizable energy is not totally available for production. Some proportion must be utilized for physical activity, keeping warm, combating disease and other stresses as well as the metabolic processes by which milk, meat, wool and offspring are produced. These costs are referred to as heat of production. What remains for maintenance and production is termed Net Energy for maintenance (NE_m), Net Energy for gain (NE_g) or Net Energy for lactation (NE_l).

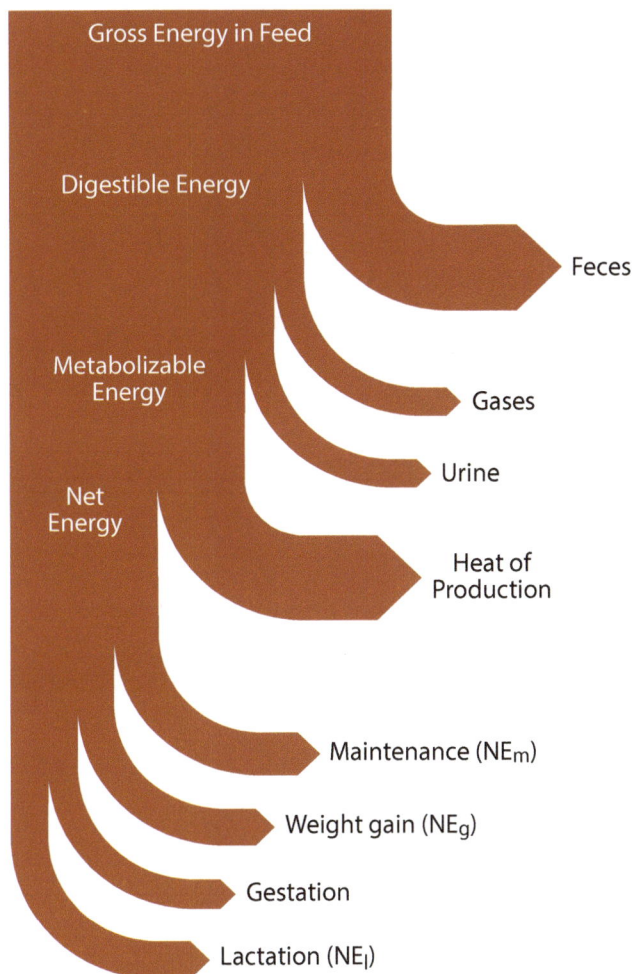

Figure 4.2: Feed energy utilization by the sheep.

For example, muscular growth may be limited by inadequate protein although adequate energy is provided and the excess of energy over protein will be stored as fat. Stored energy may be later mobilized when energy demand exceeds that available in the diet. A few numerical examples of the way in which energy is utilized are given in table 4.3.

The energy concentration in a ration affects both intake and efficiency of utilization. Figure 4.3 demonstrates that intake decreases as ME concentration in the diet increases from about 2.0 to 3.0 Mcal/kg. At ME levels below about 2.0 Mcal/kg, intake decreases as the result of increased dietary fibre which reduces the rate of digestion. As far as possible, as energy level falls, the animal consumes more in an attempt to satisfy energy requirements until the absolute limitations of bulk make this impossible.

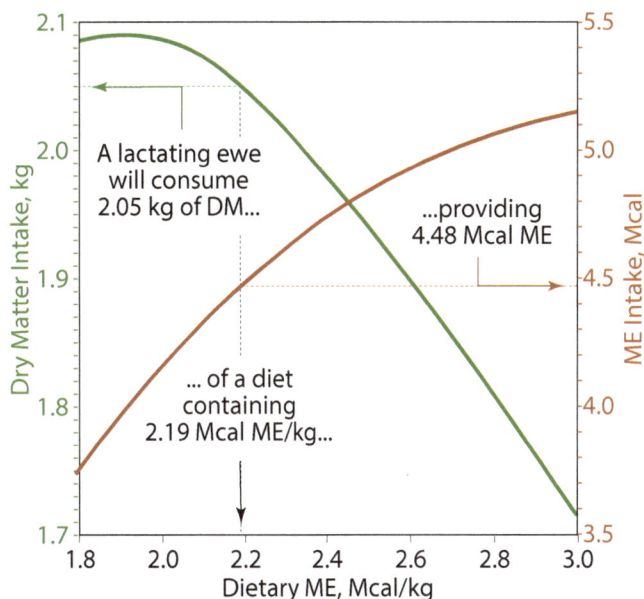

Figure 4.3: The relationship between dietary ME concentration and the intake of dry matter and ME.

Finally, energy that is available in excess of that which can be utilized for production or reproduction is stored, largely in the form of fat. This need not imply that fat is stored only after the animal has reached its maximum potential production. When rations are not balanced with respect to the nutrients they contain, fat may be stored even at lower levels of production.

Energy Fraction	20 kg lamb gaining 275 grams/day		45 kg lamb gaining 250 grams/day		70 kg ewe during maintenance		70 kg ewe 140 days preg with twins		70 kg ewe first 8 weeks lact with twins	
	kcal	%	kcal	%	kcal	%	kcal	%	kcal	%
Gross Energy Intake	4397	100	7486	100	5273	100	9259	100	12323	100
Digestible Energy	3210	73	5240	70	2900	55	5370	58	8010	65
Metabolizable Energy	2630	60	4300	57	2380	45	4400	48	6570	53
Net Energy for:										
Maintenance	1012	23	1380	18	1618	31	1618	17	1618	13
Pregnancy							330	4		
Lactation									2765	22
Gain	1040	24	1400	19						

Table 4.3: Examples illustrating the metabolic partitioning of dietary energy.

Figure 6.2: Feed wasted is profit lost.

relatively easy to waste up to 30% of the total resource. Several factors influence waste:

Physical Form: Since long hay is largely wasted when animals pull feed out of the feed bunk, chopping can reduce this problem. Chopping also limits the tendency for animals to select leaves in preference to stems. However, chopping requires a significant investment in equipment which, for most small sheep operations, would be difficult to justify.

Feed Bunk Design: The perfect, waste-free feed bunk for long hay has yet to be designed, although some are better than others. In particular those which allow the animals unlimited access to the feed with little impediment to pulling feed out are the most wasteful (figure 6.2). Designs which reduce access by using closely spaced diagonal slats or tombstone entries have been successful in preventing wastage, as have feeders like that shown in figure 6.3. Some consideration should also be given to reducing contamination of wool by feed. This can be achieved by allowing access to feed only in the lower part of the feeder.

Figure 6.3: A low waste hay feeder design.

Feeding on the ground is associated with extreme waste as well as fecal contamination leading to health problems. It can only be recommended when feed is placed on a clean area of snow where there is little risk of contamination.

Amount of Feed Offered: Although *ad libitum* feeding is recommended for growing lambs, it is usually not necessary when feeding ewes. For example, in early gestation, an 80 kg ewe may consume up to 2.5 kg of hay offered *ad lib*, while her requirements demand an intake of only 1.5 kg (9.3% CP; 2.42 Mcal DE/kg).

The situation depicted in figure 6.2 is often a result of overfeeding rather than a particularly poor feeder design. When ewes are offered more than they require, they tend to become increasingly selective. This is especially true when feeding alfalfa hay. Animals tend to eat the leaves in preference to the stems. When there is abundant feed offered, they will leave the stems behind resulting in significant waste. The solution is to feed an amount sufficient to satisfy the flock's requirements based upon knowledge of feed quality. This will serve to minimize selectivity and wastage except when the feed is very unpalatable or fibrous as is often the case with very mature, poor quality forages.

Finally, it should be realized that when feed intake is restricted (not fed *ad lib*) you will need to provide enough feed bunk space for all animals to eat at once (about 50 cm per head for ewes; 30 cm for lambs). In contrast, self-feeders require a provision of only 15 cm per head for ewes (8 cm for lambs). Nevertheless, the investment in increased feeder space is usually abundantly returned in saved feed.

Ration changes

Digestion of feed in the ruminant is dependent upon a diverse population of bacteria and protozoa in the rumen and reticulum (figure 3.4). This population adapts to each specific ration and adaptation requires time while some microbial species expand in number and others decline. Bacteria that break down cellulose are not well adapted to starch digestion. Some microbes produce mainly acetic acid from carbohydrates, others mainly propionic acid.

When rations are changed, time should be allowed for microbial adaptation. This is particularly true when grain is added to a forage ration. For example, when grain is introduced to ewes before lambing, it should be done in ¼ kg increments over the course of a week or more. This results in better feed utilization and reduced risk of digestive disturbances. The latter is often seen when animals accidentally gain access to grain storage. Sudden consumption of large quantities of grain results in the production of high levels of lactic acid in the rumen, lowering the pH and destabilizing the entire fermentation process. Lactic acid enters the bloodstream causing acidosis and bloating may ensue. Recovery may be a prolonged process.

The use of urea in sheep rations also demands a period of adaptation. The proliferation of urease-producing bacteria (p. 29) is a relatively slow process, and unless this is recognized, productivity will suffer because of the inability of the microbial population in the rumen to use urea for protein synthesis.

Mineral feeding

A significant majority of western Canadian feeds are deficient in one or several essential mineral elements. In particular, it is assumed without question that cobalt and iodine are universally in short supply, while selenium, copper (both absolutely and relative to molybdenum), zinc and phosphorus deficiencies often limit animal productivity.

There are three methods of providing mineral supplementation:

1. *Incorporation in the ration.* For example, lambs on full feed may have their mineral requirements added to a complete pelleted formulation. Alternatively, minerals may be incorporated into a supplement pellet designed for use with whole barley. When ewes are being fed grain, minerals may be added as a top dressing. These are ideal methods of assuring adequate intake, but are seldom used because of the expense of having complete rations formulated and the additional labour involved in mixing or top dressing on the farm.

2. *Provision of mineral and salt free-choice in separate containers.* This is probably the most common method of providing mineral supplementation. However, it is often unsuccessful for one of several reasons:
   - The mineral mix is unpalatable for reasons which are poorly understood; palatability and consumption of a mineral mix can vary from farm to farm and animal to animal.
   - Mineral is often allowed to become fouled by manure because it is poorly placed.
   - The mixture becomes wet, causing it to form a hard mass and promoting oxidation reactions within the mix, decreasing its feeding value.

3. *Provision of mineral in a mixture with salt.* This usually solves the palatability and consumption problem because stock will always consume salt. However, fouling and moisture can still produce problems.

In some areas, a complete mineral mix (including salt) is available which is designed to overcome the common deficiencies experienced in that area. Where such a custom mix is unavailable, the following recommendations have proven practical:

1. *Construct a mineral feeder* such as that illustrated in figure 6.4 which will:
   - be portable so that it can be moved from barn to pasture;
   - be easy to keep clean;
   - be covered to keep the mineral mix dry.

2. *Purchase a mineral mix*, preferably one designed specifically for sheep. It should have calcium and phosphorus levels in the 10–20% range and a copper level not exceeding 0.05% (see p. 45). Levels of the other minerals can vary quite widely and should rarely be of primary concern but READ THE LABEL.

3. *Purchase a trace mineralized salt with selenium.* The salt level will be in the 95% range and the selenium should be at 25 ppm (or 25 grams/tonne or 25 mg/kg). Again, the copper level must not exceed 0.1% and again READ THE LABEL.

4. *Mix thoroughly* one part of the mineral mix with one part of the trace mineralized salt with selenium. The copper level will be 0.05% at most but should be at least 0.015% to prevent deficiency. Only mix as much as the sheep will consume in 5 to 10 days, then put out fresh mix. This will promote consumption and limit the amount of moisture uptake.

5. *Provide fresh, clean water* and locate the mineral feeder close by. Poor quality or insufficient water is definite discouragement to a good appetite (p. 37).

Figure 6.4: An easy to clean, portable salt and mineral feeder.

A ewe should consume a minimum of 10 grams per day of the 1:1 mixture. This means that 20 ewes should consume two kilograms in 10 days. If more than 20 grams/ewe/day is consumed, decrease the proportion of mineral mix. Try, for example, a 1:2 mixture of mineral:TM salt. If less than 10 grams/ewe/day are consumed, increase the proportion of mineral.

These are blanket recommendations which will satisfy the flock's mineral requirements at all stages of production under most conditions of management and environment. They provide a wide margin of safety and, in doing so, will result in extra costs in many situations. Feed testing and ration formulation based on nutrient analysis is the only way to assure good levels of productivity at minimum cost.

Administering vitamins

As suggested earlier (pp. 46–47) the only vitamins normally required by sheep are A, D and E. Green, leafy forages provide adequate quantities of these but forages which have been weathered, heated or stored for prolonged periods are generally deficient.

Since all three of these 'fat-soluble' vitamins are stored in body tissues, supplementation is required only periodically, a common method of delivery being intramuscular injection. This is required only every 10 to 12 weeks during the winter feeding period. Vitamin E may also be administered in combination with selenium either:
• to ewes, two to three weeks before lambing begins; or
• to lambs, at birth,
in an attempt to eliminate the occurrence of white muscle disease in the lambs.

Vitamins are also present in most mineral mixes but these should not be considered reliable sources. In the presence of minerals and moisture, the vitamins are rapidly oxidized. Since mineral mixes rapidly take up moisture in all but the driest climates, there can be little assurance that sufficient active vitamins will remain.

In some areas, a vitamin ADE-Se premix is available which can be added to the ration or to the mineral being offered free-choice. In the latter case, care should be taken to prepare only as much premix-mineral as will be fed out in a week. Again, the mixture must be kept dry.

Grazing management

In many operations, grazed forage makes up a significant proportion of total feed intake. The efficiency with which pastures are utilized for animal production is a function of both the pasture itself and the management of the grazing animals. Figure 6.5 suggests

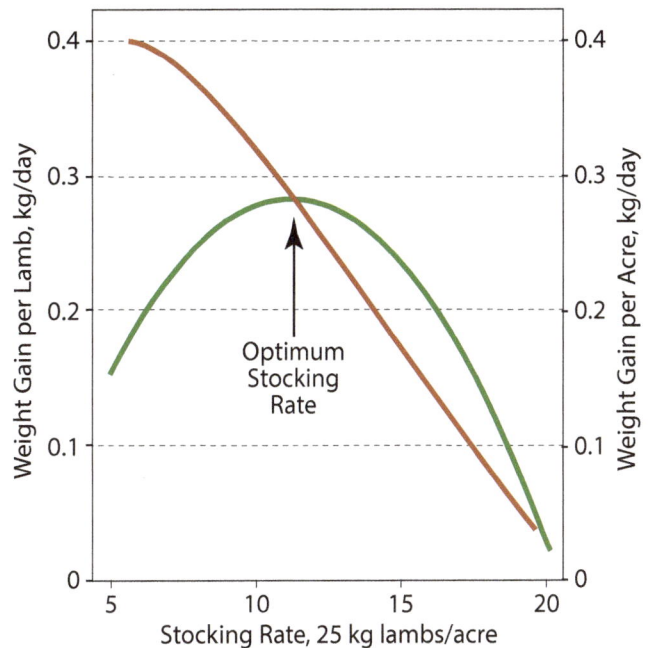

Figure 6.5: The effect of pasture stocking rate on productivity per animal versus productivity per acre.

relationships between stocking rates and both animal production per acre and productivity per animal. The main goal of grazing management is to increase animal concentration (stocking rate) to the point where animal production per unit of pasture area is maximized without significantly reducing productivity per animal. Many factors influence pasture productivity and grazing management decisions, including:
• climate;
• soil;
• fertility;
• forage species;
• nutrient requirements of livestock;
• predator control;
• and numerous others.

It is beyond the scope of this publication to discuss these in detail.

Starting newborn lambs

Starvation is the most common cause of lamb mortality and the level of loss in a flock is strictly related to management. Unless lambs receive colostrum within two or three hours of birth, body energy reserves become critically depleted. Feeding management at this time should include tipping each ewe at lambing, checking the udder, removing wax plugs in the teat ends and assisting each lamb with its first meal. This should be followed by careful scrutiny of ewes and lambs two or three times a day and providing nursing assistance until one is confident that each lamb is off to a successful start.

Colostrum should be stored in frozen form in the event that a ewe has an insufficient amount to meet the requirements of her lambs. This may be taken from a ewe that has lost her lambs or a ewe that has an obvious excess. Colostrum may also be obtained from goats or cows for emergency use. The best colostrum is obtained from older animals since, having been exposed to more diseases, they will be producing a wider range of antibodies. In addition, the level of antibodies is highest in the first colostrum taken.

Orphan lambs

The decision to raise a lamb as an orphan should be made within 24 hours of birth since the success of training to a rubber nipple decreases with advancing age. An effective procedure is as follows:

1. Allow the designated orphan to remain with its natural mother until the evening of the day of birth making certain that an adequate amount of colostrum has been consumed.
2. Late in the evening of the day of birth, isolate the lamb into a small pen (e.g., a claiming pen) preferably with solid walls and, if necessary, a heat lamp.
3. Allow the lamb to fast overnight (no more than 8 hours) and in the morning begin training to a round-holed rubber nipple on a bottle containing milk or milk replacer at near body temperature.
4. Repeat the training process at two to three hour intervals, allowing the lamb to consume no more than 50–100 ml (depending on size) each time.
5. As acceptance of the nipple progresses, the temperature of the milk can be decreased at successive feedings. If the lamb is to continue to be hand-fed, the milk temperature should ultimately reach ambient temperature (10–20 °C). If being transferred to a bucket or otherwise self-fed, training should continue until the lamb will readily accept a bottle with a crosscut nipple (p. 30) containing milk at refrigerator temperature (2–6 °C).
6. In general, small amounts of milk consumed at frequent intervals will yield better results than large volumes fed infrequently. This simulates what occurs naturally when a lamb nurses its mother.

The management of orphan self-feeding is most successful when milk (or replacer) is kept cold (2–6 °C). When a (typically 5 gallon) nipple pail is used, a block of ice in a one litre ice cream pail will keep the temperature down. Low temperature milk causes the lambs to limit their intake at each feeding. Warm milk is often consumed to the point of engorgement resulting in digestive disturbance. An additional aid in the prevention of digestive problems is the addition of 0.1% formalin to the milk.

This reduces the growth of bacterial contaminants but should not be considered a substitute for good sanitation.

When raising orphan lambs, it is particularly valuable to encourage consumption of solid feed as soon as possible. Under intensive management, orphan lambs may be weaned as early as three weeks of age minimizing labour, milk replacer costs and the risks inherent in feeding liquid diets. However, this should not be attempted unless the lambs are vigorous, free of health problems and are consuming at least 250 g/day of creep ration.

Creep feeding

The practice of encouraging early consumption of solid feed serves to promote development of a functional rumen, increasing the ability of the lamb to efficiently utilize nutrients. The primary objective of creep feeding is to provide supplemental nutrients to the lambs during the nursing period to support rapid growth. Creep feeding in confinement has little relevance when lambing is timed to coincide with the availability of spring pasture. It is most useful when lambing precedes pasture availability by six weeks or more. In particular, in areas where the grazing season is short, confinement creep feeding can give lambs a significant head start on the pasture. Creep feeding on pasture is dealt with below.

A common practice in western Canada is to lamb in late February or early March, feeding the ewes hay and grain until pasture becomes available. Lambs are either not offered creep feed or the management of the creep feeding program is such that consumption of creep ration by the lambs is minimal. Ewes and lambs are then put to pasture for the summer and at the end of the grazing season 25–50% of the lambs have not yet reached market size. They are either sold as feeders or the extra gain is put on with hay or grain.

This practice fails to take advantage of the high feed conversion efficiency (FCE) of which young lambs are capable. In fact, a pound of liveweight gain can be realized from less than two pounds of feed (FCE less than 2:1) in lambs six weeks to two months of age (table 4.13). On the other hand, lambs born in mid-March and coming off pasture in mid-September at a weight of 35–40 kg will likely need to consume about seven pounds of a 50:50 hay and grain ration for every pound of liveweight gain (FCE 7:1).

Clearly a dollar spent on grain in a creep ration is a better investment than the same dollar spent on grain in the fall. Similarly, grain is more efficiently fed to a young lamb than to its mother after two months of lactation.

Another advantage of creep feeding relates to early weaning. Lambs that are consuming more than 250 g/day of creep ration can be safely weaned allowing placement in the feedlot or separate pasture use by ewes and lambs.

Design of the creep area

Figure 6.6 demonstrates several points concerning the design of a creep area. Panel A is the most common design used but it has two inherent faults:

- Near the end of the creep feeding period, the oldest male lambs are often wider than those small ewes which have dropped in condition from heavy lactation. It may be impossible to space the pickets to allow these lambs to enter while excluding all of the ewes.
- A heavy lamb making a rapid exit from the creep can often knock down one of the pickets, leaving a wide space for the ewes to enter. If there were a significant amount of creep ration in the feeder, grain overload could result.
- Placement of a horizontal 1 x 4, to provide a 12 inch high opening with 10 inch spacing between pickets will improve panel A. The horizontal 1 x 4 will discourage ewes from entering the creep and is easily adjustable in height. If it is placed on the outside of the panel, it will become impossible for lambs to knock down single pickets.

Panel B is a good design but also invites comment:

- Placement of the three rollers is readily adjustable. While this may seem a strong feature of the design it is not necessary if a single, adjustable horizontal bar is placed across the opening.
- Access to the creep is allowed only through the roller section rather than through the entire length as in Panel A. This is handy, as it allows you to easily catch lambs in the creep by simply blocking this small exit.

Panel C is a useful innovation which allows expansion of the creep area as the lamb crop grows. A similar panel would replace Panel A. Plan on an allowance of 1.5–2.0 square feet of creep area per lamb.

The design of the creep feeder itself should include the following:

- *Limited access.* Lambs should not be able to put their feet into the feeder. Dirty feed is unattractive to them and fecal contamination spreads diseases such as coccidiosis.
- *Sufficient capacity.* Although a small quantity of fresh feed should be put out twice daily in the first few weeks, a capacity of at least 1 pound of feed per lamb should be planned for later in the creep feeding period.

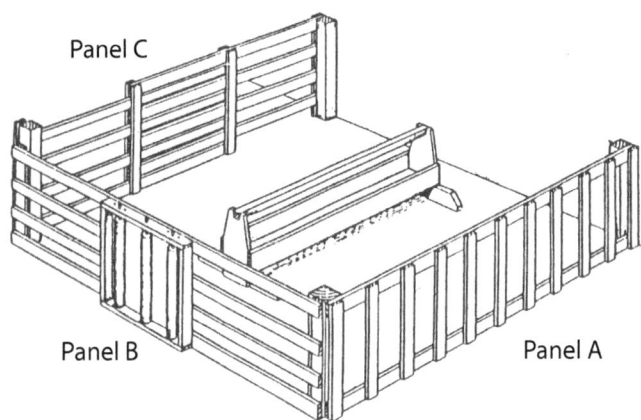

Figure 6.6: Features of a lamb creep pen.

- The feeder shown in figure 6.6 is too open. Lambs will not hesitate to stand with their front feet in the feed. Placement of a horizontal board 6" above the throat board would improve the design. An alternative is shown in figure 6.7.

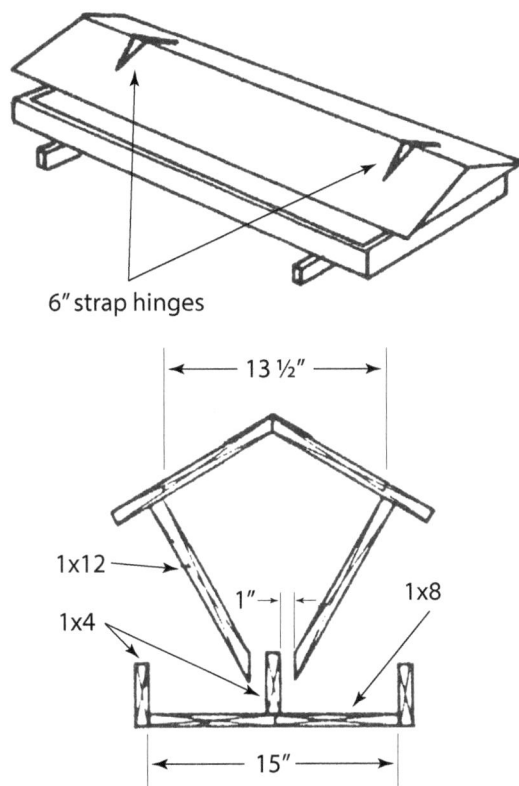

Figure 6.7: A well-designed creep self-feeder.

Fresh, clean water should be provided in the creep area once the oldest lamb is three weeks of age.

Lamb psychology

Three factors dominate the behaviour of young lambs and a well-designed creep feeding program must recognize them:

- *Attachment to the ewe.* A young lamb will not stray far from its mother. Once out of eyesight or earshot for any length of time, both ewe and lamb will attempt to re-establish contact. With this in mind, the creep area should be placed as close as possible to the area where the ewes bed down. When in the creep, the lambs should be able to see the ewes. Unless these conditions are satisfied, lambs will not be inspired to explore the creep area.
- *Desire for comfort.* Once into the creep, lambs will be encouraged to stay if the area is well bedded, warm and bright. If the environment does not dictate the use of a heat lamp, a 450 lumen light source will make the area more attractive at night.
- *Curiosity.* Lambs are inquisitive and will tend to explore their surroundings as long as it is not at the expense of their security. Curiosity will attract them into the creep and will encourage them to taste the feed offered. This behaviour can be reinforced by conspicuously adding a small amount of fresh feed twice a day for the first few weeks.

Creep ration

Many differences of opinion exist concerning appropriate ingredients and physical form of creep rations. However, the main factors to be considered are nutritional adequacy and palatability. The feeding program should be designed well in advance, and changes that might set back the lambs' progress should be avoided. The following management practices are recommended:

- As early as possible after lambing begins, set up the creep and put out a very small quantity of soybean meal. It is one of the most palatable feeds commonly available and, although it is expensive, the lambs will consume an insignificant amount. Make sure the meal is always available and always fresh. Do not put out so much that it starts to accumulate in the bottom of the feeder. If it does, refresh it daily.
- When the lambs are consuming 30–60 grams per head per day, start adding a 3:1 mixture of whole barley and pelleted 32% protein supplement (containing no urea). In most cases, the lambs will begin to demonstrate a preference for the pellets and barley and, over a period of 7–10 days the proportion of soybean meal can be gradually decreased.
- When the lambs reach a weight range of 10–15 kg change the proportion of barley to supplement from 3:1 to 5:1.
- When pasture becomes available, maintain access to the creep area and keep feed available. The lambs will gradually change themselves over to pasture consumption.

- If you plan to raise the lambs in a feedlot situation, remove the ewes at weaning time so as not to disturb the feeding patterns of the lambs. Again, maintain access to the creep area and keep feed available. The importance of fresh, uncontaminated feed cannot be overemphasized. Feeders should never be empty nor should stale feed be allowed to accumulate.

Expected results

If they can be attracted to the creep area, lambs will begin to nibble feed at two weeks of age. At six weeks of age they should be consuming close to 250 grams per day which is likely increasing their average daily gain by 0.1 kg.

By two months of age, the total creep consumption will have reached 10 kg per lamb resulting in a 5+ kg head-start on pasture or the feedlot. Figure 6.7 relates consumption of solid feed to lamb growth and the ewe's milk production.

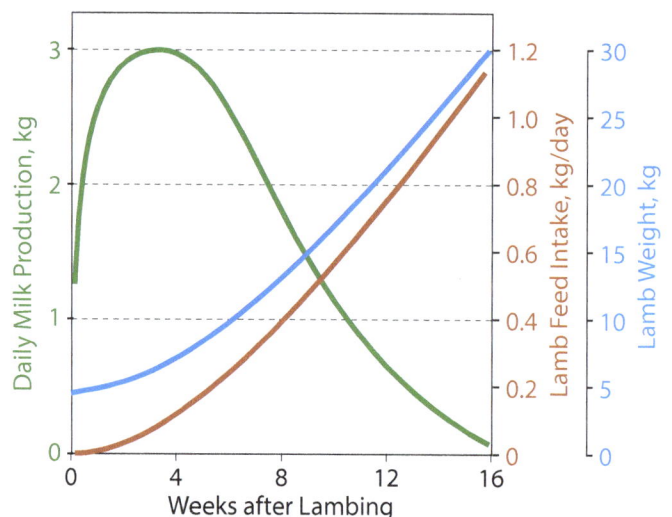

Figure 6.7: The relationship between the potential milk production of a ewe and the growth and dry feed intake of one of her twin lambs.

Creep feeding on pasture

When lambing coincides with the beginning of pasture growth in the spring, there is no advantage in encouraging the consumption of concentrates. In fact, early spring pasture is very high in nutrient content (table 2.2) and lambs tend to prefer this feed over any other. If the flock is confined at night, creep feeding of concentrates may be attempted, but usually with little success for two reasons:

- concentrate intake is generally low because pasture intake is preferred; and
- even when the concentrate is very palatable and a significant amount is consumed, this apparently does

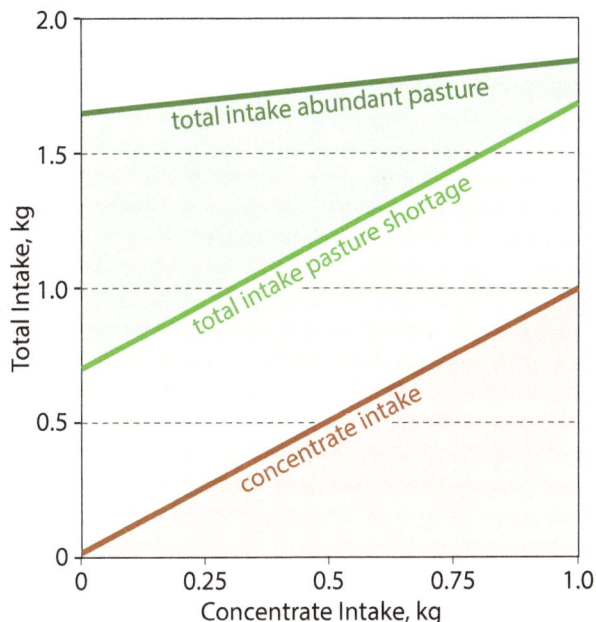

Figure 6.8: When pasture is abundant, concentrate intake replaces pasture intake. Concentrate supplements forage intake when pasture is in short supply.

not supplement pasture consumption but simply substitutes for it (figure 6.8).

- If pastures have been allowed to decline in quality the above observations may not apply. Reduced pasture palatability may result in a preference for concentrate.

Forward creep grazing

When ewes and lambs are pastured together, lambs are subject to competition from their dams for the most palatable (and highest quality) forage. This tends to discourage forage consumption by the lambs and inhibits productivity. Forward creep grazing is a technique designed to overcome this competition.

When paddock, strip or rotational grazing strategies are used, lambs are allowed access to the next (forward) grazing area before the ewes (figure 6.9). This is accomplished by setting up a creep panel between the 'present' and the 'forward' areas. The practice allows for the separate control of grazing pressures in the two areas. Grazing pressure in the 'forward' area should always be light.

Likewise, in early lactation, grazing pressure in the 'present' area should be light to promote milk production. But, as lactation progresses and milk production potential declines, grazing pressure in the 'present' area can be increased without penalizing the growth of the lambs. Increased stocking rates in the 'present' area will also tend to encourage lambs through the creep panel.

In order for forward creep grazing to be successful, lambs must be trained and much of the same psychology applies here as was described in the previous section (pp. 67–68). The following management practices are recommended:

1. Set up the creep panel as early as possible to exploit the young lambs' desire to explore.
2. Reduce the size of 'present' paddocks as much as possible to increase the lambs' chances of finding the creep panel. The use of small paddocks will, of course, necessitate more rapid rotations in order to maintain lactation at a high level.
3. Place a small, portable feeder in the 'forward' area. A spare mineral feeder (fig. 6.4) would be adequate.
4. Twice or three times a day put a small amount (e.g. 0.5 kg; 1.1 lb) of concentrate into the feeder. The act of doing so will stimulate the lambs' curiosity.
5. Once the lambs have become accustomed to using the 'forward' area, these practices can be discontinued. The discovery of choice herbage available without competition will encourage the lambs to return.

Weaning

The decision to wean or not depends on the overall flock management situation. For example, when lambs are sold to the Easter market at 20–25 kg, weaning is irrelevant. On a small farm with limited acreage, forward creep grazing may be practiced without a need for weaning.

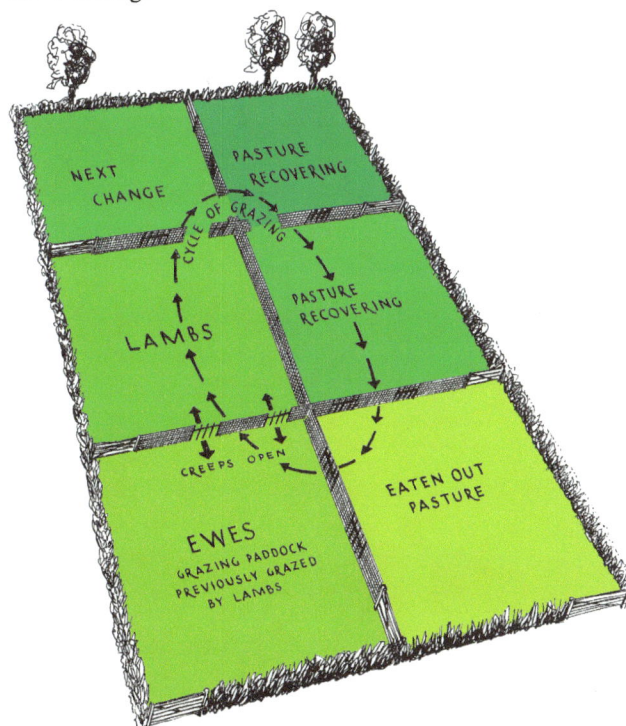

Figure 6.9: Forward creep grazing.

Management decisions around weaning should be based on the following considerations:

- In the first few weeks of life, the lamb is incapable of digesting anything but milk. However, the conversion of feed to milk by the ewe is at best only about 66% efficient. Beginning at about three weeks of age the development of the lamb's rumen makes it more efficient to encourage consumption of solid feed, bypassing the ewe. In addition, the ewe's milk production typically peaks about three weeks of age and declines steadily thereafter (figure 4.15). After about 8 weeks, the milk's contribution to the lamb's total nutrient requirement is usually quite small.
- Where pastures are varied (e.g., native range vs. cultivated pasture), weaning makes it possible to use the best quality pasture for lambs without competition.
- Some producers elect to keep lambs in confinement after weaning, using the pasture base to graze only the ewes. Lambs can be grown out quickly, free of intestinal roundworms.
- If pasture can be reserved for weaned lambs, parasite infestations are often significantly reduced since there is no contamination by older animals.
- It is commonly believed that early weaned lambs gain more uniformly since their variable reliance on the ewes is eliminated.

When to wean

It is impossible to recommend an exact age and weight for weaning. One of the most important criteria is the consumption of solid feed, reflecting the development of the rumen. A lamb must be consuming a minimum of 250 grams per day before weaning. Under intensive management this may be achieved as early as three weeks of age. When feeding expensive milk replacers to orphan lambs, there is an increased economic incentive to such very early weaning.

Researchers in Scotland have suggested that lambs weaned at about 6 weeks of age make the most efficient use of grazed feed (table 6.1). These trails also suggested that weaning at 6 weeks was less stressful than weaning at 13 or 20 weeks, but only when the consumption of solid feed was adequate.

Weaning Age, weeks	Feed Efficiency (kg feed/kg gain)
6	3.4
13	5.6
20	5.5

Table 6.1: Early weaning can yield highly efficient gains.

Weaning in confinement

Successful early weaning in confinement is absolutely dependent upon a well managed creep feeding program. This involves offering a palatable, nutritious ration and clean, fresh water in a well lighted, comfortable creep, placed where the lambs will be attracted to it. Before the time comes to wean, the following points should be considered in order to reduce lamb stress and minimize the risk of mastitis in the ewe:

1. Lambs should have been vaccinated with 7-or 8-way clostridial vaccine.
2. A week before weaning, ewes should be changed from the high quality legume hay fed during lactation to a lower quality (but palatable) grass hay.
3. Four days before weaning, grain should be removed from the ewes' ration.
4. The day before weaning, ewes should be denied access to water.
5. At weaning, ewes should be removed from the lambs so that they are at least out of sight and, preferably, out of earshot.

Weaning on pasture

Early weaning is seldom an objective when the flock is on pasture and consequently the considerations outlined above are unnecessary. However, it is often desirable to wean in mid-summer when pasture quality begins to deteriorate. Lambs may be transferred to the feedlot for finishing or they may be finished on annual pastures such as oats, pasture rape or kale. In either case, the success of weaning will depend upon the degree of nutritional independence achieved by the lambs. This is a function of pasture management. If pastures have been maintained in the early vegetative (highly palatable) stage, at weaning, the lambs will be deriving most of their nutritional requirements through grazing. Milk demand will have decreased and weaning will present few complications.

Feeding weaned lambs

In the introduction to **Chapter 1**, it was suggested that sheep and other ruminants are relatively inefficient in their conversion of feed to human food. This observation was made in the context of the animal production system as a whole. However, the potential feed conversion efficiency of the lamb itself rivals that of the feeder pig. Targets of 4 to 5 pounds of feed per pound of grain are well within reach for many of our genetically superior animals. Management practices for feeding lambs should be designed to take advantage of this potential. This is not meant to imply that such feed efficiencies can be realized in all management systems. On pasture, for example, lower conversion rates are offset by decreased feed costs. The ultimate measure of success is therefore the cost efficiency of feeding or the cost of feed per pound of gain.

Cost efficient gains

The process of lamb growth was described earlier. Lambs have the potential to grow very quickly when they are young, but as they age and the costs of maintenance increase, there is a decrease in this potential. Feed conversion efficiency (FCE) is highly correlated with average daily gain (ADG). In simple terms, rapid gains are efficient gains.

As growth progresses and FCE declines, the cost of the daily feed consumption approaches the market value of the daily gain. It is cost efficient to feed lambs to the point where the two values are equal. (figure 6.10). Of course, the actual weight at which this occurs depends upon:

- *The individual animal.* Clearly, individuals within a crop of lambs grow at widely varying rates depending upon their gender, their genetic potential, their feed intake and their health status.
- *The cost of gain.* As suggested above, when feed costs are low, slower gains can be tolerated. Neverless, given a particular feed, every effort should be made to maximize its conversion to gain. For example, it makes no sense to restrict the intake of feeder lambs (pp. 55–56). When calculating the cost of gain, consider not only the cost of feed but other factors involved in extended feeding periods, such as veterinary costs, extra labour and facilities, interest on investment and risk of death loss.
- *The market price of lamb.* This will vary with the market, the season, the market grade and the size of the carcass.

Some lambs have the ability to grow efficiently to heavy weights. In addition, many private customers will prefer the larger cuts from a heavier carcass. If the market is there, and gains are efficient, these lambs should be allowed to reach the heavier weights provided they do not become overfinished. Although extra costs are incurred in allowing these animals to continue growing, additional profit will be realized as long as the gains are cost efficient.

Feedlot lambs

Many producers prefer to raise lambs in confinement, practising creep feeding and early weaning with the pasture base being used exclusively for the ewes. There are several advantages to this approach:

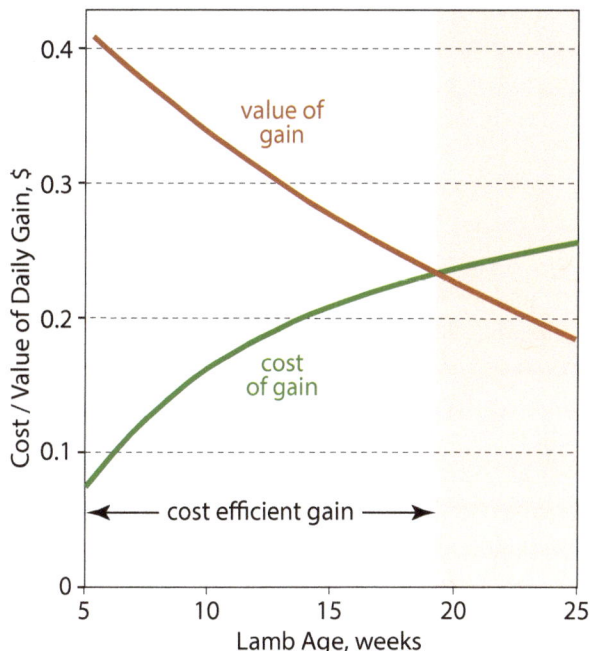

Figure 6.10: Lamb weight gains are cost efficient up to the point where the value of the daily gain is equal to the cost of producing that gain.

- Lambs can be grown out quickly on high energy rations, taking advantage of high feed conversion efficiencies and getting lambs to market before the seasonal decline in price (figure 4.21).
- The detrimental effects of gastrointestinal roundworms are minimized since these parasites require pasture to propagate.
- The loss of lambs to predators is eliminated.
- Pasture management requirements are reduced.

The disadvantages include the following:
- Unit feed cost ($/kg) are often higher since rations are usually based on supplemented grain.
- Lambs in confinement are more subjected to the spread of communicable diseases such as coccidiosis, pneumonia, sore-mouth and pinkeye.
- Investment in facilities may be higher.

Feed intake

Lambs fed for efficient gains should have *ad lib* access to feed. This implies that intake is limited only by the animals' appetite. In addition, any steps that might increase daily nutrient intake should be taken. These may include:
- Increasing ration palatability by adding flavours to the feed (e.g. molasses, anise);
- Increasing the energy content of the ration by using barley rather than oats;
- Shearing lambs being fed during hot weather; and
- Providing access to a source of clean, fresh water (p. 37). In hot weather, cold water promotes feed intake.

When feed intake is restricted, a greater proportion of the total nutrient intake is required for maintenance with a smaller contribution available for gain. The result is slower growth and reduced feed conversion efficiency (p. 48).

Processing forages for lambs

Chopping, grinding or milling hay can reduce waste and decrease the ability of the lambs to select, for example, alfalfa leaves in preference to stems. Processing reduces the digestibility of hay because its rate of passage through the rumen is increased, allowing less time for microbial degradation. At the same time, however, the rapid passage allows a greater intake of digestible nutrients. When fed in the ground form, particle sizes should be no smaller than 1–2 inches. More complete milling can produce rumen impaction and respiratory problems associated with dust.

Ground forages that have been pelleted (e.g., alfalfa pellets) have much the same feeding value as the same feeds simply ground. The same applies to cubes. Pelleting or cubing, however, may in some cases be accompanied by significant heating, resulting in heat damage to forage proteins.

Whole grains in lamb rations

Experiments conducted at the University of British Columbia on the effect of feeding whole, rolled or pelleted grains to lambs revealed higher growth rates and feed efficiency when whole grain is fed compared to the same ration pelleted or fed as a mash.

Feeding whole grain offers the following advantages:
- Feed intake may be increased by 25% while feed utilization remains the same for whole compared to pelleted grain.
- Growth rate is faster with whole grain (table 1.2).
- Feed conversion efficiency is improved 5 to 10%.
- Whole grains produce a firmer, more desirable fat finish on the carcass.
- Whole grain does not cause damage to the lining of the rumen.
- With whole grain, there is less chance of lambs going 'off feed' and of overeating disease or acidosis problems.

Much of the benefit of whole grains can be explained on the basis of better acceptance by the lambs and the higher level of feed intake. The lambs are very efficient at chewing the grains while eating and ruminating which results in them being digested just as well as preground grain. The physical form of the grain fibre remains intact in whole grain and results in slower fermentation in the rumen. Because of this, the common problems associated with high grain feeding are greatly reduced and there is, in fact, no need to provide supplemental roughage when feeding whole grain to market lambs.

In situations where forages are fed with grain there is also evidence that whole grain is preferable to pellets. Feed intake tends to be higher and the utilization of the forage is improved. The best choice of grain will vary with location, supply and price. In western Canada, barley is usually preferred for lamb feeding.

Based on these results, the use of whole barley with a 32% protein (35.5% DM basis) supplement pellet such as that described in figure 2.6 is recommended.

Nursing lambs up to 15 kg may be creep fed a mix of three parts barley to one part supplement. Initially, soya meal added to the mix may provide some additional incentive to early consumption. For lambs of 15–30 kg the mix is five parts barley to one part supplement and beyond 30 kg, a 7:1 ration is used. These rations produce levels of about 18%, 16% and 15% crude protein respectively.

Feeding behaviour

Lambs have the ability to be very selective in their feeding behaviour when they are given the opportunity. When grazing, they will select the youngest growth. If hay is fed, they will strip off the leaves and reject the stems. When whole grains and pelleted supplements are fed, care must be taken to ensure that both ingredients are consumed in proportion to the amounts offered. Lambs may select one ingredient over another on two criteria:
- *Texture.* Pellets may be very hard and difficult to break or they may be so soft that a significant proportion of the supplement becomes a mash at the bottom of the feeder.
- *Flavour.* Some pellets contain molasses, making them very palatable with the result that the lambs select them out of the grain mixture. On the other hand, pellets containing significant amounts of urea may be unpalatable with preference being shown towards the whole grain.

Ration sorting can be reduced by limiting the amount of feed in front of the lambs. When hand feeding (to appetite), it is more effective to put out small quantities of feed at frequent intervals than to put out large quantities infrequently. This may also have the effect of increasing intake. When self-feeders are used, the space through which feed is passed from bunk to the feed tray should be minimal (figure 6.11). This results in a more continuous flow of mixed feed, with less opportunity for sorting or for feed to become stale.

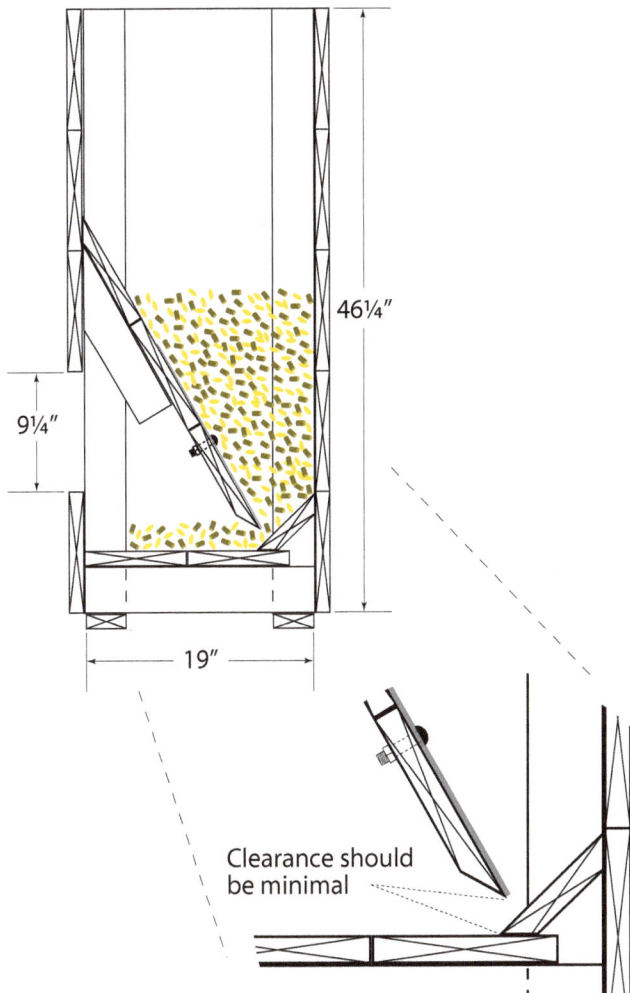

Figure 6.11: Restricting the space through which feed passes in a self-feeder minimizes feed sorting and wasting.

Slotted floors

There are several advantages to the use of slotted floors for feeding lambs:

- Space requirements are minimal, amounting to only 3–4 square feet per lamb.
- Lambs are kept cool. This is a particular advantage in the summer months when high temperatures can reduce feed intake (p. 40).
- Lambs are kept dry. In high precipitation areas, lambs may be fed under a roof of minimal size and be kept up out of the mud.
- The risk of fecal contamination leading to coccidiosis is reduced.
- No bedding is required.

Slotted floors are ineffective when more than five square feet per lamb are allowed. Higher lamb densities are required to keep manure moving down through the slots. Feeding long hay on slotted floors often results in plugging of the slots.

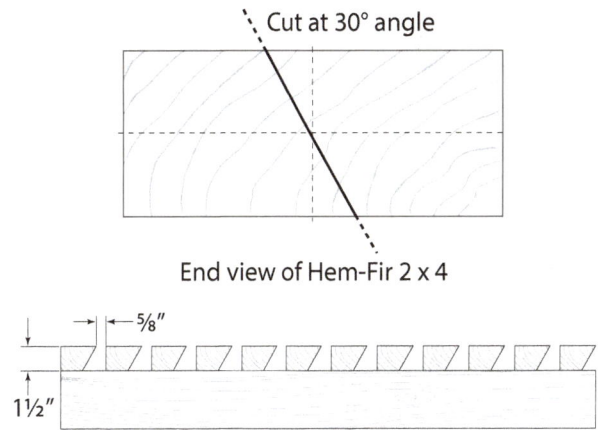

Figure 6.12: Slotted floor section.

When slotted floor sections are constructed of wood, the slats should be undercut as shown in figure 6.12. Straight-sided slots can easily become plugged.

Pasture lambs

Young, leafy pasture provides a high energy level and sufficient protein, vitamins and minerals to permit growth rates in excess of 0.3 kg/day. On the other hand, pasture which has entered the reproductive state is significantly lower in nutrient content (p. 26) and may be capable of supporting little more than maintenance.

Pasture maturity also affects intake, with immature growth being far more palatable than reproductive growth. The effects of quality and palatability in combination significantly affect animal response.

Even when intensive pasture management is practised lambs should be allowed to graze under minimum pressure. High performance per animal should not be sacrificed for production per acre at this time (figure 6.10) if maximum advantage is to be taken of the lambs' potential feed conversion efficiency.

Annual pastures

Although few producers in western Canada have incorporated annual pastures into their management systems, they have significant potential for many. Their main use would be to supply forage in late summer and fall when many perennial pastures are declining in yield and quality. In some cases they can be used to supplement grazing of hay crop regrowth.

Crops suitable for annual pastures include fall rye, forage oats, pasture rape, kale and hybrid brassicas such as Skyfall stubble turnips. Most require 6–12 weeks to reach a stage where they can be efficiently utilized. Properly managed they can provide an abundance of nutrients for lambs approaching market weight.

Shearing lambs

As stated earlier, high summer temperatures can reduce feed intake (p. 40). This effect is amplified when lambs have a significant wool covering. Heat that is produced in the digestive process cannot be dissipated because of the insulative effect of the fleece. Under these conditions, shearing may increase feed intake, daily gain and feed efficiency by as much as 10 to 15 percent. Alternatively, by using hair breeds one can avoid the practice of shearing.

Castrating ram lambs

Under good nutritional management, ram lambs grow 15–20% faster than ewe lambs with the performance of wethers being about midway between males and females. There is a significant advantage in terms of feed efficiency in leaving male lambs intact. In addition, ram lambs finish at heavier weights, often 5–10% higher than wethers. When a market for larger lambs is available, there is a definite economic advantage to adding this extra weight (p. 71).

However, early castration of male lambs also has its advantages. Unless lambs are marketed before they reach 5 months of age, intact males may become a management problem. They must be separated from the female lambs, in particular those to be selected as flock replacements. Even isolated, homosexual behaviour among intact males may reduce their interest in feed, resulting in decreasing gains.

In some instances, packers and abattoirs will apply penalties against male lambs on the basis of poor eating quality or increased skinning time. Although several research studies have examined these points, there is little evidence to suggest that they are valid reasons to penalize ram lambs.

An alternative to castration is the 'short-scrotum technique' which results in the creation of cryptorchid rams. This technique involves pushing the testicles up into the abdomen and placing an elastrator ring high up on the empty scrotum (figure 6.13). The scrotum drops off and the heat of the body within the abdomen prevents the formation of viable sperm. However, the testicles continue to produce male hormones resulting in no difference in feed efficiency in cryporchids compared with intact ram. It should be noted that this procedure is subjected to periodic failure and even a single fertile male can undo a well-planned ewe lamb replacement program.

Market lamb appraisal

It is important for the producer to be able to assess the market readiness of lambs so that the maximum return can be realized. For this purpose a scale is absolutely essential. If a large number of lambs are marketed, convenient handling facilities are also a must.

In addition to measuring their actual weights, some method of assessing the lambs' degree of finish is required. A technique similar to that used in condition scoring ewes has been developed in the UK and is equally applicable here. This method relies upon feeling the dock (tail head), loin and rib with the objective of judging how the carcass will grade after slaughter. Table 6.2 illustrates the areas to be assessed and the criteria for establishing fat class. The finish desired will depend on the market into which lambs are sold but fat classes 2–3 are generally considered ideal. The ultimate objective is to produces carcasses like those illustrated in figure 6.14.

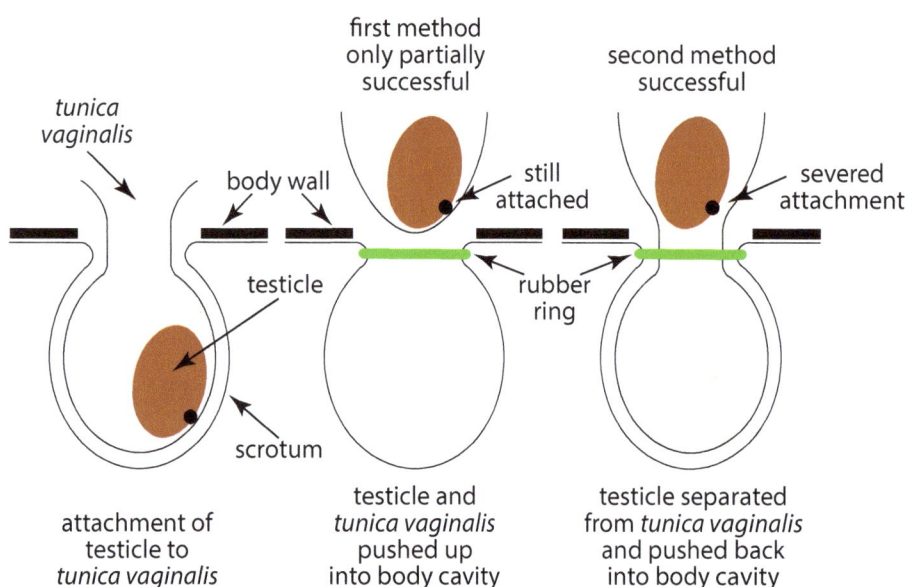

Figure 6.13: The 'short scrotum' method is a practical alternative to castration if it is correctly performed.

Fat Class	Dock	Loin	Rib
1	Individual bones very easy to detect	Very easy to feel between processes which are very prominent	Individual processes can be easily felt
2	Individual bones easy to detect with light pressure	Prominent spinous and transverse processes felt easily	Individual ribs show slight cover but still easy to detect
3	Moderate pressure required to detect individual bones	Tips of processes rounded. Individual bones felt as corrugations with light pressure	Individual ribs have softer feel, with fat cover becoming more evident in between and over ribs, which are now less easy to detect
4	Firm pressure required to detect individual bones	Spinous processes felt with moderate pressure; transverse processes felt with firm pressure	Individual ribs are only detectable with firm pressure
5	Individual bones cannot be detected	Individual processes cannot be felt	Individual ribs are undetectable, soft, rolling, spongy feel

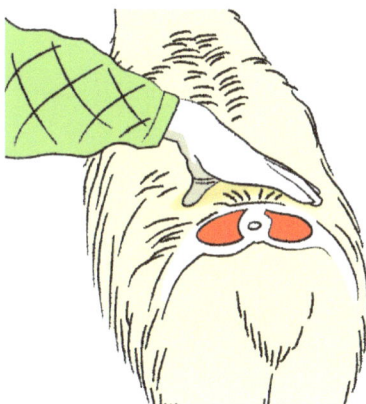

Table 6.2: Market lamb appraisal criteria. source: UK Agriculture and Horticulture Development Board 2019, Understanding lambs & carcases for better returns.

Figure 6.14: Market lamb carcasses ideally finished.

Appendix A: Nutrient Requirement Tables

The nutrient requirements tabulated here are derived from two primary sources:

- Requirements for energy, protein, calcium and phosphorus are derived from Nutrient Requirements of Sheep 1985, published by the (US) National Research Council (NRC).
- Requirements for minerals other than calcium and phosphorus are derived from the NRC Nutrient Requirements of Small Ruminants 2007. This publication is primarily based on defining requirements across a broad range of grazing animals (sheep, goats, deer, alpacas and llamas); it does not deal effectively with energy, protein, calcium and phosphorus requirements for sheep raised under conditions typical for western Canadian sheep enterprises.

Requirements for immature animals are based on the theoretical growth curves illustrated in figure A1. Average weights for mature animals, provided by the Canadian Sheep Breeders Association are as follows:

- small breeds (e.g., Border Cheviot, Romanov, Shetland, Icelandic): ewes – 60 kg; rams – 85 kg;
- medium-sized breeds (e.g., Dorset, Dorper, Texel, Ile de France, Katahdin, Rideau, Canadian, Outaouais, Rambouillet): ewes – 80 kg; rams – 105 kg;
- large breeds (e.g., Suffolk, Hampshire, Charollais, Columbia, Polypay); ewes –100 kg; rams – 140 kg.

Rates of gain are derived from 2018 Canadian lamb growth data provided by the GenOvis performance testing program.

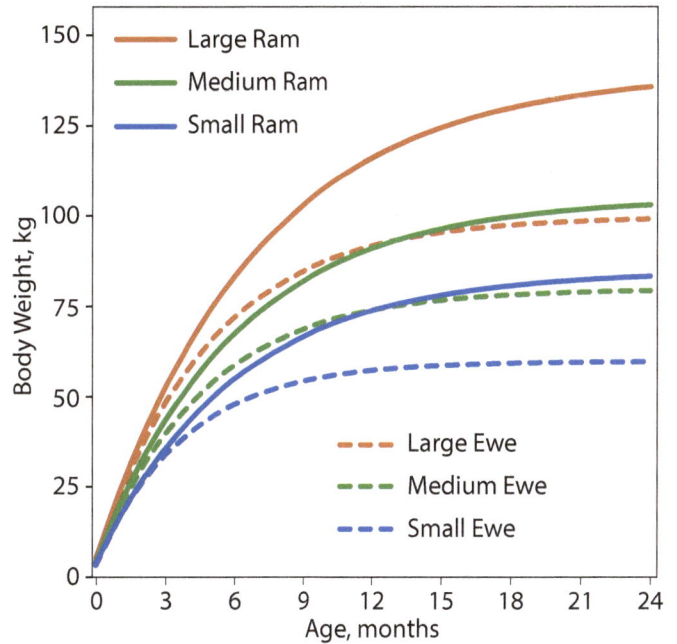

Figure A1. Standard reference growth curves for sheep of varying mature sizes.

Body Weight kg	Daily Gain grams	DM Intake kg	LBW/ Milk[1] kg	Digestible Energy Mcal/day	Mcal/kg	MACRONUTRIENT REQUIREMENTS Crude Protein g/day	% in DM	Calcium g/day	% in DM	Phosphorus g/day	% in DM
YEARLING AND MATURE EWES											
Maintenance											
50	10	1.0		2.42	2.42	94	9.4	2.0	0.20	2.0	0.20
60	10	1.1		2.66	2.42	103	9.4	2.2	0.20	2.2	0.20
70	10	1.2		2.90	2.42	113	9.4	2.4	0.20	2.4	0.20
80	10	1.3		3.15	2.42	122	9.4	2.6	0.20	2.6	0.20
90	10	1.4		3.39	2.42	132	9.4	2.8	0.20	2.8	0.20
100	10	1.5		3.63	2.42	141	9.4	3.0	0.20	3.0	0.20
110	10	1.6		3.87	2.42	150	9.4	3.2	0.20	3.2	0.20
Flushing – 2 weeks prebreeding and first 3 weeks of breeding											
50	100	1.6		4.15	2.60	146	9.1	5.1	0.32	2.9	0.18
60	100	1.7		4.41	2.60	155	9.1	5.4	0.32	3.1	0.18
70	100	1.8		4.67	2.60	164	9.1	5.8	0.32	3.2	0.18
80	100	1.9		4.93	2.60	173	9.1	6.1	0.32	3.4	0.18
90	100	2.0		5.19	2.60	182	9.1	6.4	0.32	3.6	0.18
100	100	2.1		5.45	2.60	191	9.1	6.7	0.32	3.8	0.18
110	100	2.2		5.71	2.60	200	9.1	7.0	0.32	4.0	0.18
Nonlactating – 3-15 weeks of gestation											
50	30	1.2		2.90	2.42	112	9.3	3.0	0.25	2.4	0.20
60	30	1.3		3.15	2.42	121	9.3	3.3	0.25	2.6	0.20
70	30	1.4		3.39	2.42	130	9.3	3.5	0.25	2.8	0.20
80	30	1.5		3.63	2.42	140	9.3	3.8	0.25	3.0	0.20
90	30	1.6		3.87	2.42	149	9.3	4.0	0.25	3.2	0.20
100	30	1.7		4.11	2.42	158	9.3	4.3	0.25	3.4	0.20
110	30	1.8		4.36	2.42	167	9.3	4.5	0.25	3.6	0.20
Last 4 weeks of gestation (130-200% lambing rate expected)											
50	180	1.6	5.0	4.15	2.60	171	10.7	5.6	0.35	3.7	0.23
60	180	1.7	5.5	4.41	2.60	182	10.7	6.0	0.35	3.9	0.23
70	180	1.8	6.0	4.67	2.60	193	10.7	6.3	0.35	4.1	0.23
80	180	1.9	6.5	4.93	2.60	203	10.7	6.7	0.35	4.4	0.23
90	180	2.0	7.0	5.19	2.60	214	10.7	7.0	0.35	4.6	0.23
100	180	2.1	7.5	5.45	2.60	225	10.7	7.4	0.35	4.8	0.23
110	180	2.2	8.0	5.71	2.60	235	10.7	7.7	0.35	5.1	0.23
Last 4 weeks of gestation (200-300% lambing rate expected)											
50	225	1.7	6.0	4.86	2.86	192	11.3	6.8	0.40	4.1	0.24
60	225	1.8	6.5	5.15	2.86	203	11.3	7.2	0.40	4.3	0.24
70	225	1.9	7.0	5.43	2.86	215	11.3	7.6	0.40	4.6	0.24
80	225	2.0	7.5	5.72	2.86	226	11.3	8.0	0.40	4.8	0.24
90	225	2.1	8.0	6.01	2.86	237	11.3	8.4	0.40	5.0	0.24
100	225	2.2	8.5	6.29	2.86	249	11.3	8.8	0.40	5.3	0.24
110	225	2.3	9.0	6.58	2.86	260	11.3	9.2	0.40	5.5	0.24

Table A1a: Recommended ration macronutrient concentrations for ewes during maintenance, breeding and gestation.

Body Weight kg	Daily Gain grams	DM Intake kg	LBW/Milk[1] kg	Na	Cl	K	Mg	S	Co	Cu	Fe	I	Mn	Mo	Se	Zn
				--------------- % in DM ---------------					------------------------- mg/kg DM -------------------------							
YEARLING AND MATURE EWES																
Maintenance																
50	10	1.0		0.06	0.05	0.50	0.09	0.18	0.15	4.0	9.9	0.5	14.0	0.5	0.038	27
60	10	1.1		0.07	0.05	0.52	0.10	0.18	0.15	4.3	10.3	0.5	15.1	0.5	0.040	29
70	10	1.2		0.07	0.06	0.54	0.10	0.18	0.15	4.5	10.6	0.5	16.1	0.5	0.041	31
80	10	1.3		0.07	0.06	0.55	0.11	0.18	0.15	4.6	10.8	0.5	16.9	0.5	0.042	32
90	10	1.4		0.08	0.06	0.56	0.12	0.18	0.15	4.8	11.1	0.5	17.6	0.5	0.043	34
100	10	1.5		0.08	0.06	0.57	0.12	0.18	0.15	4.9	11.3	0.5	18.2	0.5	0.044	35
110	10	1.6		0.08	0.06	0.58	0.12	0.18	0.15	5.0	11.4	0.5	18.7	0.5	0.045	36
Flushing – 2 weeks prebreeding and first 3 weeks of breeding																
50	100	1.6		0.04	0.03	0.43	0.07	0.18	0.15	2.5	22.5	0.5	12.3	0.5	0.086	26
60	100	1.7		0.05	0.04	0.45	0.08	0.18	0.15	2.8	22.0	0.5	13.1	0.5	0.084	27
70	100	1.8		0.05	0.04	0.46	0.08	0.18	0.15	3.0	21.5	0.5	13.9	0.5	0.082	29
80	100	1.9		0.06	0.04	0.48	0.09	0.18	0.15	3.2	21.1	0.5	14.5	0.5	0.081	30
90	100	2.0		0.06	0.05	0.49	0.09	0.18	0.15	3.3	20.8	0.5	15.1	0.5	0.080	31
100	100	2.1		0.06	0.05	0.50	0.10	0.18	0.15	3.5	20.5	0.5	15.7	0.5	0.078	32
110	100	2.2		0.06	0.05	0.51	0.10	0.18	0.15	3.6	20.2	0.5	16.2	0.5	0.077	33
Nonlactating – 3-17 weeks gestation																
50	30	1.2		0.07	0.05	0.49	0.11	0.18	0.15	8.9	13.1	0.5	12.7	0.5	0.050	25
60	30	1.3		0.07	0.06	0.51	0.12	0.18	0.15	8.7	13.1	0.5	13.8	0.5	0.051	27
70	30	1.4		0.08	0.06	0.52	0.12	0.18	0.15	8.6	13.2	0.5	14.7	0.5	0.051	29
80	30	1.5		0.08	0.06	0.54	0.13	0.18	0.15	8.5	13.3	0.5	15.5	0.5	0.051	30
90	30	1.6		0.08	0.06	0.55	0.13	0.18	0.15	8.3	13.3	0.5	16.2	0.5	0.052	32
100	30	1.7		0.08	0.06	0.56	0.13	0.18	0.15	8.2	13.3	0.5	16.8	0.5	0.052	33
110	30	1.8		0.08	0.07	0.56	0.14	0.18	0.15	8.2	13.4	0.5	17.3	0.5	0.052	34
Last 4 weeks gestation (130-200% lambing rate expected)																
50	180	1.6	5.0	0.06	0.04	0.45	0.11	0.18	0.15	5.1	20.0	0.5	16.7	0.5	0.034	42
60	180	1.7	5.5	0.06	0.05	0.47	0.11	0.18	0.15	5.5	21.1	0.5	18.0	0.5	0.037	43
70	180	1.8	6.0	0.06	0.05	0.48	0.12	0.18	0.15	5.8	22.1	0.5	19.3	0.5	0.040	44
80	180	1.9	6.5	0.07	0.05	0.49	0.12	0.18	0.15	6.0	23.0	0.5	20.4	0.5	0.042	45
90	180	2.0	7.0	0.07	0.05	0.51	0.13	0.18	0.15	6.3	23.8	0.5	21.3	0.5	0.044	46
100	180	2.1	7.5	0.07	0.06	0.52	0.13	0.18	0.15	6.5	24.5	0.5	22.2	0.5	0.046	47
110	180	2.2	8.0	0.07	0.06	0.53	0.14	0.18	0.15	6.7	25.2	0.5	23.0	0.5	0.047	48
Last 4 weeks gestation (200-300% lambing rate expected)																
50	225	1.7	6.0	0.06	0.04	0.45	0.11	0.18	0.15	5.3	21.8	0.5	17.3	0.5	0.036	45
60	225	1.8	6.5	0.06	0.05	0.46	0.12	0.18	0.15	5.6	22.7	0.5	18.5	0.5	0.038	46
70	225	1.9	7.0	0.06	0.05	0.48	0.12	0.18	0.15	5.9	23.6	0.5	19.6	0.5	0.040	47
80	225	2.0	7.5	0.07	0.05	0.49	0.13	0.18	0.15	6.1	24.4	0.5	20.7	0.5	0.043	48
90	225	2.1	8.0	0.07	0.05	0.50	0.13	0.18	0.15	6.4	25.0	0.5	21.6	0.5	0.044	48
100	225	2.2	8.5	0.07	0.06	0.51	0.13	0.18	0.15	6.6	25.7	0.5	22.4	0.5	0.046	49
110	225	2.3	9.0	0.07	0.06	0.52	0.14	0.18	0.15	6.7	26.3	0.5	23.2	0.5	0.048	50

Table A1b: Recommended ration mineral concentrations for ewes during maintenance, breeding and gestation.

Body Weight kg	Daily Gain grams	DM Intake kg	LBW/ Milk[1] kg	Digestible Energy Mcal/day	Mcal/kg	Crude Protein g/day	% in DM	Calcium g/day	% in DM	Phosphorus g/day	% in DM
YEARLING AND MATURE EWES (continued)											
First 6-8 weeks of lactation suckling singles											
50	-25	2.1	0.80	6.01	2.86	281	13.4	6.7	0.32	5.5	0.26
60	-25	2.3	0.90	6.58	2.86	308	13.4	7.4	0.32	6.0	0.26
70	-25	2.5	1.00	7.15	2.86	335	13.4	8.0	0.32	6.5	0.26
80	-25	2.6	1.10	7.44	2.86	348	13.4	8.3	0.32	6.8	0.26
90	-25	2.7	1.20	7.72	2.86	362	13.4	8.6	0.32	7.0	0.26
100	-25	2.8	1.30	8.01	2.86	375	13.4	9.0	0.32	7.3	0.26
110	-25	2.9	1.40	8.29	2.86	389	13.4	9.3	0.32	7.5	0.26
First 6-8 weeks of lactation suckling twins											
50	-60	2.4	1.40	6.86	2.86	360	15.0	9.4	0.39	7.0	0.29
60	-60	2.6	1.55	7.44	2.86	390	15.0	10.1	0.39	7.5	0.29
70	-60	2.8	1.70	8.01	2.86	420	15.0	10.9	0.39	8.1	0.29
80	-60	3.0	1.85	8.58	2.86	450	15.0	11.7	0.39	8.7	0.29
90	-60	3.2	2.00	9.15	2.86	480	15.0	12.5	0.39	9.3	0.29
100	-60	3.4	2.15	9.72	2.86	510	15.0	13.3	0.39	9.9	0.29
110	-60	3.6	2.30	10.30	2.86	540	15.0	14.0	0.39	10.4	0.29
Last 4-6 weeks of lactation suckling singles											
50	45	1.6	0.24	4.15	2.60	171	10.7	5.6	0.35	3.7	0.23
60	45	1.7	0.28	4.41	2.60	182	10.7	6.0	0.35	3.9	0.23
70	45	1.8	0.32	4.67	2.60	193	10.7	6.3	0.35	4.1	0.23
80	45	1.9	0.36	4.93	2.60	203	10.7	6.7	0.35	4.4	0.23
90	45	2.0	0.40	5.19	2.60	214	10.7	7.0	0.35	4.6	0.23
100	45	2.1	0.44	5.45	2.60	225	10.7	7.4	0.35	4.8	0.23
110	45	2.2	0.48	5.71	2.60	235	10.7	7.7	0.35	5.1	0.23
Last 4-6 weeks of lactation suckling twins											
50	90	2.1	0.40	6.01	2.86	281	13.4	6.7	0.32	5.5	0.26
60	90	2.3	0.45	6.58	2.86	308	13.4	7.4	0.32	6.0	0.26
70	90	2.5	0.50	7.15	2.86	335	13.4	8.0	0.32	6.5	0.26
80	90	2.6	0.55	7.44	2.86	348	13.4	8.3	0.32	6.8	0.26
90	90	2.7	0.60	7.72	2.86	362	13.4	8.6	0.32	7.0	0.26
100	90	2.8	0.65	8.01	2.86	375	13.4	9.0	0.32	7.3	0.26
110	90	2.9	0.70	8.29	2.86	389	13.4	9.3	0.32	7.5	0.26

Table A2a: Recommended ration macronutrient concentrations for ewes during lactation.

Body Weight kg	Daily Gain grams	DM Intake kg	LBW/ Milk[1] kg	OTHER MACROMINERALS					MICRO (TRACE) MINERALS							
				Na	Cl	K	Mg	S	Co	Cu	Fe	I	Mn	Mo	Se	Zn
				---------------- % in DM ----------------					-------------------------- mg/kg DM --------------------------							

YEARLING AND MATURE EWES (continued)

First 6-8 weeks lactation suckling singles

50	-25	2.1	0.80	0.05	0.07	0.47	0.08	0.18	0.15	3.0	6.8	0.8	9.1	0.5	0.130	29
60	-25	2.3	0.90	0.05	0.07	0.49	0.09	0.18	0.15	3.2	7.2	0.8	9.8	0.5	0.135	31
70	-25	2.5	1.00	0.05	0.08	0.50	0.09	0.18	0.15	3.3	7.5	0.8	10.4	0.5	0.138	32
80	-25	2.6	1.10	0.06	0.08	0.51	0.10	0.18	0.15	3.6	8.1	0.8	11.3	0.5	0.147	35
90	-25	2.7	1.20	0.06	0.09	0.53	0.10	0.18	0.15	3.8	8.7	0.8	12.1	0.5	0.155	37
100	-25	2.8	1.30	0.06	0.09	0.54	0.11	0.18	0.15	4.0	9.2	0.8	12.9	0.5	0.162	40
110	-25	2.9	1.40	0.07	0.10	0.56	0.12	0.18	0.15	4.2	9.7	0.8	13.7	0.5	0.169	42

First 6-8 weeks lactation suckling twins

50	-60	2.4	1.40	0.05	0.09	0.51	0.10	0.18	0.15	2.9	8.2	0.8	9.8	0.5	0.191	35
60	-60	2.6	1.55	0.05	0.10	0.52	0.10	0.18	0.15	3.1	8.6	0.8	10.5	0.5	0.196	37
70	-60	2.8	1.70	0.06	0.10	0.53	0.10	0.18	0.15	3.2	9.0	0.8	11.1	0.5	0.201	39
80	-60	3.0	1.85	0.06	0.10	0.54	0.11	0.18	0.15	3.4	9.3	0.8	11.6	0.5	0.204	41
90	-60	3.2	2.00	0.06	0.11	0.55	0.11	0.18	0.15	3.5	9.6	0.8	12.1	0.5	0.208	42
100	-60	3.4	2.15	0.06	0.11	0.55	0.12	0.18	0.15	3.6	9.8	0.8	12.5	0.5	0.211	43
110	-60	3.6	2.30	0.06	0.11	0.56	0.12	0.18	0.15	3.7	10.0	0.8	12.8	0.5	0.213	44

Last 4-6 weeks lactation suckling singles

50	45	1.6	0.24	0.05	0.05	0.46	0.08	0.18	0.15	3.5	5.7	0.8	9.4	0.5	0.063	28
60	45	1.7	0.28	0.05	0.06	0.48	0.09	0.18	0.15	3.9	6.4	0.8	10.6	0.5	0.070	30
70	45	1.8	0.32	0.06	0.06	0.50	0.09	0.18	0.15	4.2	7.0	0.8	11.7	0.5	0.076	32
80	45	1.9	0.36	0.06	0.07	0.51	0.10	0.18	0.15	4.4	7.6	0.8	12.6	0.5	0.081	34
90	45	2.0	0.40	0.06	0.07	0.53	0.10	0.18	0.15	4.7	8.1	0.8	13.5	0.5	0.086	36
100	45	2.1	0.44	0.07	0.07	0.54	0.11	0.18	0.15	4.9	8.6	0.8	14.2	0.5	0.091	38
110	45	2.2	0.48	0.07	0.08	0.55	0.11	0.18	0.15	5.1	9.0	0.8	14.9	0.5	0.095	39

Last 4-6 weeks lactation suckling twins

50	90	2.1	0.40	0.04	0.05	0.44	0.07	0.18	0.15	2.8	5.0	0.8	7.7	0.5	0.072	28
60	90	2.3	0.45	0.04	0.05	0.45	0.08	0.18	0.15	2.9	5.4	0.8	8.4	0.5	0.075	29
70	90	2.5	0.50	0.05	0.06	0.46	0.08	0.18	0.15	3.1	5.7	0.8	8.9	0.5	0.077	30
80	90	2.6	0.55	0.05	0.06	0.47	0.08	0.18	0.15	3.3	6.2	0.8	9.8	0.5	0.082	32
90	90	2.7	0.60	0.05	0.06	0.49	0.09	0.18	0.15	3.5	6.7	0.8	10.5	0.5	0.087	33
100	90	2.8	0.65	0.06	0.07	0.50	0.09	0.18	0.15	3.8	7.1	0.8	11.2	0.5	0.091	35
110	90	2.9	0.70	0.06	0.07	0.51	0.10	0.18	0.15	4.0	7.5	0.8	11.9	0.5	0.095	36

Table A2b: Recommended ration mineral concentrations for ewes during lactation.

Body Weight kg	Daily Gain grams	DM Intake kg	LBW/Milk[1] kg	Digestible Energy Mcal/day	Digestible Energy Mcal/kg	Crude Protein g/day	Crude Protein % in DM	Calcium g/day	Calcium % in DM	Phosphorus g/day	Phosphorus % in DM
EWE LAMBS											
Non-lactating - first 15 weeks of gestation											
40	160	1.4		3.63	2.60	148	10.6	4.9	0.35	3.1	0.22
50	135	1.5		3.89	2.60	159	10.6	5.3	0.35	3.3	0.22
60	135	1.6		4.15	2.60	170	10.6	5.6	0.35	3.5	0.22
70	125	1.7		4.41	2.60	180	10.6	6.0	0.35	3.7	0.22
80	125	1.8		4.67	2.60	191	10.6	6.3	0.35	4.0	0.22
90	125	1.9		4.93	2.60	201	10.6	6.7	0.35	4.2	0.22
Last 4 weeks of gestation (100-120% lambing rate expected)											
40	180	1.5	4.0	4.16	2.77	177	11.8	5.9	0.39	3.3	0.22
50	160	1.6	4.5	4.44	2.77	189	11.8	6.2	0.39	3.5	0.22
60	160	1.7	5.0	4.71	2.77	201	11.8	6.6	0.39	3.7	0.22
70	150	1.8	5.5	4.99	2.77	212	11.8	7.0	0.39	4.0	0.22
80	150	1.8	6.0	4.99	2.77	212	11.8	7.0	0.39	4.0	0.22
90	150	1.9	6.5	5.27	2.77	224	11.8	7.4	0.39	4.2	0.22
Last 4 weeks of gestation (130-200% lambing rate expected)											
40	225	1.5	5.0	4.36	2.90	192	12.8	7.2	0.48	3.8	0.25
50	225	1.6	5.5	4.65	2.90	205	12.8	7.7	0.48	4.0	0.25
60	225	1.7	6.0	4.94	2.90	218	12.8	8.2	0.48	4.3	0.25
70	215	1.8	6.5	5.23	2.90	230	12.8	8.6	0.48	4.5	0.25
80	215	1.9	7.0	5.52	2.90	243	12.8	9.1	0.48	4.8	0.25
90	215	2.0	7.5	5.81	2.90	256	12.8	9.6	0.48	5.0	0.25
First 6-8 weeks of lactation suckling singles (wean by 8 weeks)											
40	-50	1.7	0.6	4.94	2.90	223	13.1	5.1	0.30	3.7	0.22
50	-50	2.1	0.7	6.10	2.90	275	13.1	6.3	0.30	4.6	0.22
60	-50	2.3	0.8	6.68	2.90	301	13.1	6.9	0.30	5.1	0.22
70	-50	2.5	0.9	7.26	2.90	328	13.1	7.5	0.30	5.5	0.22
80	-50	2.7	1.0	7.84	2.90	354	13.1	8.1	0.30	5.9	0.22
90	-50	2.9	1.1	8.42	2.90	380	13.1	8.7	0.30	6.4	0.22
First 6-8 weeks of lactation suckling twins (wean by 8 weeks)											
40	-100	2.1	1.0	6.38	3.04	288	13.7	7.8	0.37	5.5	0.26
50	-100	2.3	1.2	6.98	3.04	315	13.7	8.5	0.37	6.0	0.26
60	-100	2.5	1.4	7.59	3.04	343	13.7	9.3	0.37	6.5	0.26
70	-100	2.7	1.6	8.20	3.04	370	13.7	10.0	0.37	7.0	0.26
80	-100	2.9	1.8	8.80	3.04	397	13.7	10.7	0.37	7.5	0.26
90	-100	3.1	2.0	9.41	3.04	425	13.7	11.5	0.37	8.1	0.26

Table A3a: Recommended ration macronutrient concentrations for ewe lambs during gestation and lactation.

Body Weight kg	Daily Gain grams	DM Intake kg	LBW/ Milk[1] kg	Na	Cl	K	Mg	S	Co	Cu	Fe	I	Mn	Mo	Se	Zn
				--------------- % in DM ---------------					------------------------- mg/kg DM -------------------------							

EWE LAMBS

Non-lactating - 3-17 weeks gestation

Body Weight kg	Daily Gain grams	DM Intake kg	LBW/ Milk kg	Na	Cl	K	Mg	S	Co	Cu	Fe	I	Mn	Mo	Se	Zn
40	160	1.4		0.06	0.02	0.45	0.10	0.18	0.15	6.6	37.1	0.5	14.8	0.5	0.141	33
50	135	1.5		0.06	0.02	0.46	0.11	0.18	0.15	6.6	30.7	0.5	14.5	0.5	0.117	31
60	135	1.6		0.06	0.02	0.48	0.11	0.18	0.15	6.6	29.7	0.5	15.3	0.5	0.113	33
70	125	1.7		0.07	0.02	0.49	0.11	0.18	0.15	6.6	27.0	0.5	15.6	0.5	0.103	33
80	125	1.8		0.07	0.02	0.50	0.12	0.18	0.15	6.6	26.3	0.5	16.2	0.5	0.101	34
90	125	1.9		0.07	0.02	0.51	0.12	0.18	0.15	6.6	25.7	0.5	16.8	0.5	0.098	35

Last 4 weeks gestation (100-120% lambing rate expected)

Body Weight kg	Daily Gain grams	DM Intake kg	LBW/ Milk kg	Na	Cl	K	Mg	S	Co	Cu	Fe	I	Mn	Mo	Se	Zn
40	180	1.5	4.0	0.05	0.01	0.43	0.10	0.18	0.15	4.5	17.1	0.5	14.2	0.5	0.147	39
50	160	1.6	4.5	0.06	0.01	0.45	0.10	0.18	0.15	4.9	18.4	0.5	15.8	0.5	0.125	39
60	160	1.7	5.0	0.06	0.02	0.46	0.11	0.18	0.15	5.2	19.6	0.5	17.3	0.5	0.123	40
70	150	1.8	5.5	0.06	0.02	0.48	0.11	0.18	0.15	5.5	20.7	0.5	18.5	0.5	0.114	41
80	150	1.8	6.0	0.07	0.02	0.50	0.12	0.18	0.15	6.1	22.9	0.5	20.7	0.5	0.119	44
90	150	1.9	6.5	0.07	0.02	0.51	0.13	0.18	0.15	6.4	23.7	0.5	21.8	0.5	0.117	45

Last 4 weeks gestation (130-200% lambing rate expected)

Body Weight kg	Daily Gain grams	DM Intake kg	LBW/ Milk kg	Na	Cl	K	Mg	S	Co	Cu	Fe	I	Mn	Mo	Se	Zn
40	225	1.5	5.0	0.06	0.01	0.44	0.11	0.18	0.15	5.0	20.4	0.5	16.0	0.5	0.165	46
50	225	1.6	5.5	0.06	0.02	0.46	0.12	0.18	0.15	5.4	21.6	0.5	17.5	0.5	0.146	47
60	225	1.7	6.0	0.06	0.02	0.47	0.12	0.18	0.15	5.7	22.6	0.5	18.8	0.5	0.142	48
70	215	1.8	6.5	0.07	0.02	0.49	0.13	0.18	0.15	6.0	23.5	0.5	20.0	0.5	0.133	48
80	215	1.9	7.0	0.07	0.02	0.50	0.13	0.18	0.15	6.2	24.3	0.5	21.1	0.5	0.130	49
90	215	2.0	7.5	0.07	0.02	0.51	0.13	0.18	0.15	6.5	25.1	0.5	22.0	0.5	0.128	49

First 6-8 weeks lactation suckling singles (wean by 8 weeks)

Body Weight kg	Daily Gain grams	DM Intake kg	LBW/ Milk kg	Na	Cl	K	Mg	S	Co	Cu	Fe	I	Mn	Mo	Se	Zn
40	-50	1.7	0.6	0.04	0.02	0.47	0.05	0.18	0.15	3.0	6.5	0.8	8.9	0.5	0.122	29
50	-50	2.1	0.7	0.04	0.02	0.46	0.05	0.18	0.15	2.9	6.3	0.8	8.8	0.5	0.116	29
60	-50	2.3	0.8	0.05	0.02	0.48	0.05	0.18	0.15	3.1	6.8	0.8	9.5	0.5	0.121	30
70	-50	2.5	0.9	0.05	0.03	0.49	0.05	0.18	0.15	3.3	7.2	0.8	10.1	0.5	0.126	32
80	-50	2.7	1.0	0.05	0.03	0.50	0.06	0.18	0.15	3.4	7.5	0.8	10.6	0.5	0.130	33
90	-50	2.9	1.1	0.05	0.03	0.50	0.06	0.18	0.15	3.5	7.8	0.8	11.1	0.5	0.134	34

First 6-8 weeks lactation suckling twins (wean by 8 weeks)

Body Weight kg	Daily Gain grams	DM Intake kg	LBW/ Milk kg	Na	Cl	K	Mg	S	Co	Cu	Fe	I	Mn	Mo	Se	Zn
40	-100	2.1	1.0	0.04	0.03	0.48	0.04	0.18	0.15	2.7	7.0	0.8	8.6	0.5	0.157	33
50	-100	2.3	1.2	0.05	0.03	0.50	0.04	0.18	0.15	2.9	7.7	0.8	9.6	0.5	0.172	37
60	-100	2.5	1.4	0.05	0.03	0.51	0.05	0.18	0.15	3.1	8.4	0.8	10.5	0.5	0.185	40
70	-100	2.7	1.6	0.06	0.03	0.53	0.05	0.18	0.15	3.3	9.0	0.8	11.3	0.5	0.197	42
80	-100	2.9	1.8	0.06	0.03	0.54	0.05	0.18	0.15	3.5	9.4	0.8	11.9	0.5	0.206	45
90	-100	3.1	2.0	0.06	0.04	0.55	0.06	0.18	0.15	3.6	9.9	0.8	12.5	0.5	0.214	47

Table A3b: Recommended ration mineral concentrations for ewe lambs during gestation and lactation.

Body Weight kg	Daily Gain grams	DM Intake kg	Digestible Energy Mcal/day	Mcal/kg	Crude Protein g/day	% in DM	Calcium g/day	% in DM	Phosphorus g/day	% in DM
REPLACEMENT EWE LAMBS[1]										
30	470	1.2	3.43	2.86	185	15.4	6.4	0.53	2.6	0.22
40	400	1.4	4.00	2.86	176	12.6	5.9	0.42	2.6	0.18
50	340	1.5	3.89	2.60	136	9.1	4.8	0.32	2.4	0.17
60	270	1.5	3.89	2.60	134	8.9	4.7	0.31	2.5	0.17
70	200	1.5	3.89	2.60	132	8.8	4.6	0.30	2.8	0.17
80	130	1.6	3.84	2.40	130	8.1	4.5	0.28	2.4	0.15
90	60	1.6	3.84	2.40	128	8.0	4.5	0.28	2.4	0.15
REPLACEMENT RAM LAMBS										
40	470	1.8	4.99	2.77	243	13.5	7.8	0.43	3.7	0.21
60	380	2.4	6.65	2.77	264	11.0	8.4	0.35	4.2	0.18
80	290	2.8	7.76	2.77	269	9.6	8.5	0.30	4.6	0.16
100	190	3.0	8.32	2.77	264	8.8	8.2	0.30	4.8	0.16
120	100	3.0	8.31	2.77	240	8.0	8.0	0.27	5.0	0.16
LAMBS FINISHING: 4-7 MONTHS OLD (medium-sized ewe)										
30	350	1.3	4.12	3.17	191	14.7	6.6	0.51	3.2	0.24
40	280	1.6	5.35	3.34	186	11.6	6.6	0.42	3.3	0.21
50	210	1.6	5.42	3.39	160	10.0	5.6	0.35	3.0	0.19
60	140	1.7	5.76	3.39	153	9.0	5.0	0.29	2.8	0.16
EARLY WEANED LAMBS - MODERATE GROWTH POTENTIAL (medium-sized ram)										
10	510	0.5	1.76	3.52	127	25.4	4.0	0.82	1.9	0.38
20	460	1.0	3.43	3.43	167	16.7	5.4	0.54	2.5	0.24
30	400	1.3	4.46	3.43	191	14.7	6.7	0.51	3.2	0.24
40	350	1.5	5.15	3.43	202	13.5	7.7	0.55	3.9	0.28
50	300	1.6	5.49	3.43	193	12.1	7.0	0.55	3.8	0.28
60	240	1.7	5.83	3.43	183	10.8	6.8	0.40	3.6	0.21
EARLY WEANED LAMBS - RAPID GROWTH POTENTIAL (large ram)										
10	620	0.6	2.11	3.52	157	26.2	4.9	0.82	2.2	0.38
20	570	1.2	4.12	3.43	203	16.9	6.5	0.54	2.9	0.24
30	520	1.4	4.80	3.43	211	15.1	7.2	0.51	3.4	0.24
40	470	1.5	5.15	3.43	218	14.5	8.6	0.55	4.3	0.28
50	430	1.7	5.83	3.43	247	14.5	9.4	0.55	4.8	0.28
60	380	1.8	6.18	3.43	261	14.5	8.2	0.55	4.5	0.28

[1] Recommendations for ewe lambs with expected mature body weight = 100 kg. For ewe lambs with lower or higher expected mature body weights, scale requirements in proportion to body weight. For example, for ewes with expected mature body weight of 80 kg, multiply recommendations by 0.8.

Table A4a: Recommended ration macronutrient concentrations for growing ewe and ram replacement lambs and mixed gender market lambs.

Body Weight kg	Daily Gain grams	DM Intake kg	OTHER MACROMINERALS					MICRO (TRACE) MINERALS							
			Na	Cl	K	Mg	S	Co	Cu	Fe	I	Mn	Mo	Se	Zn
			---------------- % in DM ----------------					---------------------------- mg/kg DM ----------------------------							
REPLACEMENT EWE LAMBS															
30	470	1.2	0.08	0.06	0.51	0.14	0.18	0.15	9.2	113	0.50	25	0.50	0.43	8.0
40	400	1.4	0.07	0.05	0.50	0.12	0.18	0.15	7.4	83	0.50	18	0.50	0.31	11.2
50	340	1.5	0.07	0.05	0.52	0.11	0.18	0.15	6.7	66	0.50	14	0.50	0.25	12.9
60	270	1.5	0.07	0.05	0.55	0.11	0.18	0.15	6.3	52	0.50	11	0.50	0.20	15.4
70	200	1.5	0.07	0.06	0.57	0.11	0.18	0.15	5.9	39	0.50	8	0.50	0.15	17.9
80	130	1.6	0.07	0.05	0.58	0.11	0.18	0.15	5.2	24	0.50	5	0.50	0.09	19.1
90	60	1.6	0.07	0.06	0.61	0.11	0.18	0.15	4.8	11	0.50	2	0.50	0.04	21.4
REPLACEMENT RAM LAMBS															
40	470	1.8	0.06	0.05	0.46	0.10	0.18	0.15	6.5	76	0.50	16	0.50	0.29	8.8
60	380	2.4	0.05	0.04	0.46	0.08	0.18	0.15	4.7	46	0.50	10	0.50	0.17	9.7
80	290	2.8	0.05	0.04	0.47	0.08	0.18	0.15	4.0	30	0.50	6	0.50	0.11	11.0
100	190	3.0	0.05	0.04	0.49	0.07	0.18	0.15	3.6	18	0.50	4	0.50	0.07	12.7
120	100	3.0	0.05	0.04	0.52	0.08	0.18	0.15	3.5	10	0.50	2	0.50	0.04	15.2
LAMBS FINISHING: 4-7 MONTHS OLD (medium-sized ewe)															
30	350	1.3	0.06	0.05	0.47	0.11	0.18	0.15	6.8	78	0.50	17	0.50	0.30	7.3
40	280	1.6	0.05	0.04	0.46	0.09	0.18	0.15	5.2	51	0.50	11	0.50	0.19	9.7
50	210	1.6	0.05	0.04	0.49	0.09	0.18	0.15	4.8	38	0.50	8	0.50	0.14	12.0
60	140	1.7	0.05	0.04	0.50	0.08	0.18	0.15	4.2	24	0.50	5	0.50	0.09	13.5
EARLY WEANED LAMBS - MODERATE GROWTH POTENTIAL (medium-sized ram)															
10	510	0.5	0.15	0.11	0.60	0.28	0.18	0.15	20.7	295	0.50	64	0.50	1.12	3.2
20	460	1.0	0.08	0.06	0.49	0.15	0.18	0.15	10.1	133	0.50	29	0.50	0.51	6.5
30	400	1.3	0.06	0.05	0.48	0.11	0.18	0.15	7.5	89	0.50	19	0.50	0.34	7.3
40	350	1.5	0.06	0.05	0.48	0.10	0.18	0.15	6.4	68	0.50	15	0.50	0.26	10.4
50	300	1.6	0.06	0.05	0.50	0.10	0.18	0.15	5.8	54	0.50	12	0.50	0.21	12.1
60	240	1.7	0.06	0.05	0.51	0.10	0.18	0.15	5.3	41	0.50	9	0.50	0.16	13.6
EARLY WEANED LAMBS - RAPID GROWTH POTENTIAL (large ram)															
10	620	0.6	0.14	0.11	0.59	0.28	0.18	0.15	20.5	299	0.50	65	0.50	1.14	2.8
20	570	1.2	0.08	0.06	0.48	0.14	0.18	0.15	10.1	138	0.50	30	0.50	0.52	5.5
30	520	1.4	0.07	0.05	0.48	0.13	0.18	0.15	8.5	108	0.50	23	0.50	0.41	6.9
40	470	1.5	0.07	0.05	0.50	0.12	0.18	0.15	7.8	91	0.50	20	0.50	0.34	10.5
50	430	1.7	0.07	0.05	0.50	0.11	0.18	0.15	6.8	73	0.50	16	0.50	0.28	11.5
60	380	1.8	0.07	0.05	0.52	0.11	0.18	0.15	6.3	61	0.50	13	0.50	0.23	12.9

Table A4b: Recommended ration mineral concentrations for growing ewe and ram replacement lambs and mixed gender market lambs.

Appendix B: Feed Library

The following tables contain typical analysis results for western Canadian feeds. Sources are as follows:

- Values in tables 1 to 5 were originally derived from the Alberta Agriculture publication 'Ten Year Average Analyses of Alberta Feeds, 1984-1994'. The immediate source of the values in those tables is the on-line SheepBytes Ration Balancer which was also originally produced by Alberta Agriculture.
- Table 6 values were compiled from advertised guaranteed analyses of manufactured feeds available in western Canada at the time of writing.
- Table 7 is a replica of the table originally published in 'Nutrition Guide for BC Sheep Producers'.
- Values in tables 8 to 12 are derived from 2 publications produced by Manitoba Agriculture in cooperation with the Manitoba Forage Council: 'Manitoba Average Feed Values for the Beef/Bison/Sheep Producer' and 'Manitoba Average Forage Mineral Concentrations'.

A series of annual summaries published by the Saskatchewan Feed Testing Service is available on-line at: www.agromedia.ca/sheep. All four western provinces operated feed testing labs up until the late 1980s or early 1990s. All of these labs have been subsequently closed.

Other summaries of feed analysis results are readily available, including:

- all of the US National Research Council (NRC) publications in the Animal Nutrition Series, including 'Nutrient Requirements of Small Ruminants', 'Nutrient Requirements of Beef Cattle' and 'Nutrient Requirements of Dairy Cattle'.
- the US National Animal Nutrition Program (NANP) maintains a comprehensive feed analysis database, accessible at: https://animalnutrition.org/user/login?destination=feed-composition-database

As explained in **Chapter 2** there are many factors that can affect the concentrations of feed fractions, especially in forages. The use of 'book values' as the basis for formulating rations is no substitute for proper sampling and analysis of the specific ingredients to be fed.

Feed Name	DM %	DE Mcal/kg	CP %	NDF %	eNDF % of NDF	DIP % of CP	UIP % of CP	Ca %	P %	Mg %	K %	S %	Na %	Cl %
ALFALFA-GRASS HAY	87.4	2.63	14.0	52.0	95.0	78.0	22.0	1.22	0.19	0.26	1.65	0.17	0.01	
ALFALFA HAY EARLY BL	87.9	2.71	18.2	47.0	92.0	81.0	19.0	1.52	0.24	0.33	1.72	0.24	0.01	
ALFALFA HAY LATE BL	87.9	2.46	12.5	56.0	92.0	81.0	19.0	1.40	0.22	0.33	1.72	0.24	0.03	
BARLEY GREENFEED	85.9	2.58	11.8	58.0	92.0	70.0	30.0	0.41	0.22	0.23	1.83	0.27	0.10	
BROME HAY	89.8	2.54	10.6	57.7	98.0	77.0	23.0	0.46	0.17	0.17	1.50	0.14	0.01	
CANOLA HAY	85.0	2.64	15.1	45.7	92.0	77.0	23.0	1.16	0.27	0.39	1.64	0.55	0.10	
CHICKPEA HAY	85.7	2.55	14.1	52.0	92.0	78.0	22.0	1.22	0.28	0.36	2.20	0.14	0.01	
CLOVER-GRASS HAY	86.2	2.61	10.8	56.0	95.0	76.0	24.0	0.79	0.19	0.17	1.62	0.14	0.01	
GERMAN MILLET	85.0	2.51	14.1	57.7	98.0	77.0	23.0	0.32	0.19	0.57	4.74	0.22	0.05	
GRASS HAY	89.9	2.55	10.7	62.8	92.0	77.0	23.0	0.53	0.17	0.17	1.32	0.18		
GRASS-LEGUME HAY	87.2	2.60	12.5	60.0	94.0	75.0	25.0	0.80	0.16	0.18	1.57	0.16	0.02	
KOCHIA HAY	88.8	2.60	14.3	47.0	92.0	81.0	19.0	0.96	0.25	0.96	3.61	0.24	0.30	
LENTIL HAY	85.7	2.60	14.2	52.0	92.0	78.0	22.0	1.28	0.23	0.36	1.28	0.14	0.01	
NATIVE GRASS HAY	91.0	2.47	8.6	72.5	92.0	75.0	25.0	0.44	0.12	0.14	1.25	0.19	0.10	
OAT GREENFEED	85.8	2.57	9.9	61.0	98.0	70.0	30.0	0.31	0.20	0.26	1.96	0.19	0.10	
ORCHARD GRASS HAY	88.4	2.67	12.9	59.6	98.0	70.0	30.0	0.46	0.21	0.17	1.87	0.11	0.01	
OVERWINTERED GRASS	60.0	2.38	5.6	69.0	41.0	63.0	37.0	0.36	0.09	0.14	1.25	0.24		
PEA HAY	85.7	2.46	14.0	52.0	92.0	78.0	22.0	1.30	0.21	0.20	1.28	0.14	0.01	
RED CLOVER HAY	86.2	2.51	15.3	52.0	92.0	78.0	22.0	1.38	0.21	0.17	1.58	0.14		
RYEGRASS HAY	85.7	2.55	11.5	61.0	98.0	65.0	35.0	0.47	0.22	0.19	1.62	0.16		
SLOUGH HAY	90.0	2.34	7.2	72.5	92.0	75.0	25.0	0.44	0.12	0.14	1.27	0.19	0.10	
SWEET CLOVER HAY	86.2	2.51	15.3	52.0	92.0	78.0	22.0	1.38	0.21	0.17	1.58	0.14		
TIMOTHY HAY	88.2	2.62	8.8	66.3	98.0	70.0	30.0	0.49	0.16	0.16	1.41	0.14	0.03	

Table B1a: Typical nutrient analyses of Alberta hay samples.

Feed Name	DM %	DE Mcal/kg	CP %	NDF %	eNDF % of NDF	DIP % of CP	UIP % of CP	Ca %	P %	Mg %	K %	S %	Na %	Cl %
ALFALFA-GRASS SILAGE	43.0	2.66	14.6	52.0	85.0	76.0	24.0	1.32	0.23	0.19	1.62	0.12	0.02	
ALFALFA SILAGE	44.6	2.75	18.2	47.0	82.0	90.0	10.0	1.77	0.25	0.27	1.82	0.27	0.01	
BARLEY SILAGE	36.8	2.75	11.1	49.0	65.0	86.0	14.0	0.46	0.27	0.22	1.60	0.22	0.20	
CANOLA SILAGE EARLY	35.0	2.48	16.2	45.7	92.0	77.0	23.0	1.29	0.33	0.45	2.16	0.55	0.08	
CANOLA SILAGE LATE	35.0	2.24	11.0	57.0	92.0	77.0	23.0	1.00	0.38	0.45	2.16	0.55	0.05	
CLOVER-GRASS SILAGE	38.9	2.58	13.9	54.0	82.0	76.0	24.0	1.08	0.23	0.21	1.44	0.10	0.03	
CORN SILAGE	35.0	2.81	10.0	50.0	81.0	80.0	20.0	0.28	0.26	0.24	1.42	0.07	0.03	
CORN SILAGE + NPN	34.0	2.90	12.0	55.0	81.0	90.0	10.0	0.32	0.28	0.24	1.42	0.07	0.03	
GRASS HAYLAGE	45.0	2.46	12.2	62.0	92.0	77.0	23.0	0.90	0.22	0.23	2.27	0.18	0.03	
LEGUME HAYLAGE	45.0	2.60	18.0	50.0	49.0	65.0	35.0	1.40	0.28	0.28	2.37	0.23	0.01	
MIXED HAYLAGE	45.0	2.55	15.8	60.0	81.0	75.0	25.0	1.21	0.27	0.27	2.31	0.15	0.02	
OAT SILAGE	37.9	2.66	10.6	58.1	61.0	85.0	15.0	0.40	0.24	0.26	1.74	0.18	0.10	
PEA SILAGE	29.5	2.89	12.3	47.0	82.0	90.0	10.0	0.49	0.32	0.21	2.10	0.27	0.05	
SWEET CLOVER SILAGE	36.5	2.54	14.8	52.0	82.0	76.0	24.0	1.40	0.22	0.27	1.59	0.19	0.03	
TRITICALE SILAGE	39.7	2.60	10.3	55.0	61.0	78.0	22.0	0.30	0.23	0.14	1.41	0.08	0.10	
WHEAT SILAGE	43.0	2.72	11.2	61.0	61.0	80.0	20.0	0.26	0.23	0.19	1.58	0.16	0.10	

Table B2a: Typical nutrient analyses of Alberta silage samples.

Feed Name	Cu mg/kg	Mn mg/kg	Zn mg/kg	Se mg/kg	I mg/kg	Co mg/kg	Fe mg/kg	Mo mg/kg	Vit A KIU/kg	Vit D IU/kg	Vit E IU/kg
ALFALFA-GRASS HAY	4	40	23	0.03			200	2			
ALFALFA HAY EARLY BL	4	40	23	0.03			213	3			
ALFALFA HAY LATE BL	4	40	23	0.03			213	3			
BARLEY GREENFEED	4	35	29	0.03			250	2			
BROME HAY	3	45	20	0.03			167	2			
CANOLA HAY	3	45	25	0.03			175	2			
CHICKPEA HAY	5	46	29	0.03			128	2			
CLOVER-GRASS HAY	5	35	26	0.03			200	2			
GERMAN MILLET	4	45	20	0.03			167	2			
GRASS HAY	4	75	24	0.03			233	1			
GRASS-LEGUME HAY	5	44	24	0.03			200	1			
KOCHIA HAY	3	40	23	0.03			213	1			
LENTIL HAY	4	40	23	0.03			128	2			
NATIVE GRASS HAY	5	33	20	0.03			167	2			
OAT GREENFEED	5	33	20	0.03			178	2			
ORCHARD GRASS HAY	2	53	26	0.03			168	2			
OVERWINTERED GRASS	4	33	20	0.03			161	2			
PEA HAY	5	46	29	0.03			128	2			
RED CLOVER HAY	4	46	29	0.03			128	2			
RYEGRASS HAY	4	32	28	0.03			281	1			
SLOUGH HAY	5	33	20	0.03			167	2			
SWEET CLOVER HAY	4	46	29	0.03			128	2			
TIMOTHY HAY	3	45	28	0.03			173	2			

Table B1b: Typical nutrient analyses of Alberta hay samples.

Feed Name	Cu mg/kg	Mn mg/kg	Zn mg/kg	Se mg/kg	I mg/kg	Co mg/kg	Fe mg/kg	Mo mg/kg	Vit A KIU/kg	Vit D IU/kg	Vit E IU/kg
ALFALFA-GRASS SILAGE	5	28	19	0.03			200	2			
ALFALFA SILAGE	4	27	22	0.03			355	2			
BARLEY SILAGE	5	33	27	0.03			300	2			
CANOLA SILAGE EARLY	3	45	25	0.03			175	2			
CANOLA SILAGE LATE	3	45	25	0.03			175	2			
CLOVER-GRASS SILAGE	5	31	21	0.03			200	2			
CORN SILAGE	7	42	42	0.03			249	2			
CORN SILAGE + NPN	7	42	42	0.03			249	2			
GRASS HAYLAGE	5	65	26	0.03			125	1			
LEGUME HAYLAGE	6			0.03			255	1			
MIXED HAYLAGE	5	37	24	0.03			145	1			
OAT SILAGE	5	40	26	0.03			173	2			
PEA SILAGE	4	27	22	0.03			355	2			
SWEET CLOVER SILAGE	8	27	27	0.03			300	1			
TRITICALE SILAGE	3	35	28	0.03			287	2			
WHEAT SILAGE	5	35	23	0.03			168	2			

Table B2b: Typical nutrient analyses of Alberta silage samples.

Feed Name	DM %	DE Mcal/kg	CP %	NDF %	eNDF % of NDF	DIP % of CP	UIP % of CP	Ca %	P %	Mg %	K %	S %	Na %	Cl %
ALFALFA SUNCURED	90.0	2.73	17.6	42.0	76.0	70.0	30.0	1.35	0.20	0.27	2.00	0.21	0.03	
ALFALFA MEAL	92.0	2.68	18.9	54.0	7.0	50.0	50.0	1.37	0.25	0.22	1.90	0.18	0.05	
APPLE POMACE - WET	22.0	3.04	5.4	42.5	34.0	100.0	0.0	0.23	0.11		0.53	0.11		
BAKERY WASTE	92.0	3.96	9.0	18.0	5.0	75.6	24.4	0.15	0.24	0.18	0.43	0.02	1.12	
BEET MOLASSES	78.0	3.31	8.5	0.0	0.0	90.0	10.0	0.15	0.03	0.29	6.06	0.59	1.50	
BEET PULP	91.0	3.29	8.8	44.6	33.0	45.0	55.0	0.66	0.11	0.30	0.23	0.23	0.21	
BREWERS GRAIN	92.0	2.90	28.1	42.0	18.0	40.0	60.0	0.19	0.68	0.18	0.27	0.85	1.00	
CANOLA MEAL	91.9	3.08	38.3	27.0	23.0	68.0	32.0	0.75	1.26	0.62	1.31	1.16	0.03	
CANOLA CAKE 10%	94.0	3.67	38.3	27.0	23.0	68.0	32.0	0.64	1.06	0.52	1.25	0.85	0.11	
CANOLA CAKE 8%	94.0	3.55	38.3	21.2	23.0	68.0	32.0	0.64	1.06	0.52	1.25	0.85	0.11	
CARROTS	12.0	3.61	10.1	0.0	0.0	100.0	0.0	0.37	0.32	0.20	2.80			
CORN & COB MEAL	86.0	3.70	9.4	0.0	0.0	100.0	0.0	0.08	0.30	0.11	0.41			
CORN DISTILLERS - DRY	92.0	3.70	30.5	38.1	4.0	45.0	55.0	0.05	0.81	0.26	0.70	0.70	0.60	
CORN GLUTEN FEED	90.0	3.52	23.8	36.2	36.0	75.0	25.0	0.07	0.95	0.40	1.40	0.47	0.26	
CORN GLUTEN MEAL	91.0	3.92	66.3	8.9	36.0	41.0	59.0	0.07	0.61	0.15	0.48	0.90	0.06	
COTTONSEED - WHOLE	93.0	4.14	23.0	40.0	100.0	69.0	31.0	0.15	0.62	0.35	1.22	0.26	0.03	
DDGS CORN	91.0	3.70	28.5	44.0	4.0	45.0	55.0	0.16	0.75	0.26	0.70	0.38	0.57	
DDGS CORN-WHEAT	92.4	3.40	33.9	42.1	4.0	48.0	52.0	0.11	0.97	0.32	0.89	0.58	0.40	
DDGS WHEAT	92.4	3.33	39.3	38.3	4.0	50.0	50.0	0.17	0.96	0.38	1.08	0.44	0.20	
DG WET WHEAT	35.0	3.33	24.0	55.0	4.0	50.0	50.0	0.34	0.60	0.22	0.55	0.40	0.20	
DISTILLERS G SYRUP	30.0	3.61	40.0	29.0	4.0	50.0	50.0	0.22	0.97	0.38	1.08	0.48	0.20	
GRAIN SCR PELLETS	90.0	3.15	12.2	0.0	6.0	70.0	30.0	0.20	0.78	0.17	0.33	0.14	0.01	
GREENFEED PELLETS	93.7	2.81	14.0	28.0	42.0	76.0	24.0	0.78	0.32	0.38	2.06	0.43		
HM CORN & COB	65.0	3.70	9.4	0.0	0.0	100.0	0.0	0.08	0.30	0.11	0.46			
LENTIL SCREENINGS	88.2	3.31	21.2	14.0	5.0	78.0	22.0	0.31	0.45	0.14	1.04	0.22	0.01	
LINSEED MEAL	90.0	3.21	28.5	25.0	23.0	65.0	35.0	0.43	0.89	0.58	1.30	0.43	0.15	
MOLASSES - WET	75.0	3.26	5.0	0.0	73.0	100.0	0.0	0.15	0.03	0.29	6.06	0.60	1.48	
MOLASSES - DRY	97.0	3.26	5.0	0.0	73.0	100.0	0.0	0.15	0.03	0.29	6.06	0.60	1.48	
PEA SCREENINGS	90.0	3.08	21.1	14.0	0.0	78.0	22.0	0.10	0.43	0.14	1.04	0.22	0.01	
POTATO WASTE - WET	14.0	3.61	9.0	0.0	0.0	80.0	20.0	0.11	0.26					
POTATOES	25.0	3.52	8.2	14.0	0.0	78.0	22.0	0.04	0.24	0.14	2.17	0.09		
SOYBEAN MEAL - 44%	89.0	3.70	49.9	14.9	23.0	65.0	35.0	0.40	0.71	0.31	2.22	0.46	0.04	
SOYBEAN MEAL - 48%	90.0	3.83	54.0	8.0	23.0	65.0	35.0	0.29	0.71	0.33	2.36	0.48	0.01	
SUNFLOWER SEEDS	90.0	2.44	30.0	0.0	0.0	73.0	27.0	0.35	0.95	0.60	1.11	0.42		
THIN STILLAGE	7.0	3.87	48.0	10.0	4.0	50.0	50.0	0.40	1.20	0.52	1.65	0.96	0.60	
TOMATO POMACE - DRY	92.0	2.95	12.6	0.0	0.0	100.0	0.0	0.43	0.59	0.18	3.34			
TURNIPS (TUBER)	9.0	3.70	12.0	0.0	0.0	100.0	0.0	0.70	0.34	0.22	2.99			
UREA	96.0		281.0	0.0	0.0	100.0	0.0							
WHEAT BRAN	88.0	3.10	17.0	0.0	0.0	100.0	0.0	0.18	1.25	0.64	1.53			
WHEAT SHORTS	90.0	3.76	20.0	0.0	0.0	100.0	0.0	0.17	0.98	0.46	1.22			
WHOLE STILLAGE	14.0	3.65	40.0	25.0	4.0	50.0	50.0	0.30	0.95	0.40	1.20	0.70	0.30	

Table B3a: Typical nutrient analyses of western Canadian by-product samples.

Feed Name	Cu mg/kg	Mn mg/kg	Zn mg/kg	Se mg/kg	I mg/kg	Co mg/kg	Fe mg/kg	Mo mg/kg	Vit A KIU/kg	Vit D IU/kg	Vit E IU/kg
ALFALFA SUNCURED	7	37	21	0.03			213	3			
ALFALFA MEAL	7	37	25	0.03			175	3			
APPLE POMACE - WET											
BAKERY WASTE	12	71	20			1.3	180				
BEET MOLASSES	22	5	18	0.03			87				
BEET PULP	14	39	1	0.03			330	1			
BREWERS GRAIN	6	32	61	0.45			200	1			
CANOLA MEAL	9	58	97	0.60			200	3			
CANOLA CAKE 10%	6	50	70	1.06	0.7		145	3			16
CANOLA CAKE 8%	6	58	64	1.20	0.7		145	3			14
CARROTS											
CORN & COB MEAL	2										
CORN DISTILLERS - DRY	6	25	89	0.42	0.1	1.0	233	2			
CORN GLUTEN FEED	7	22	73	0.30	0.1	0.1	226	3			
CORN GLUTEN MEAL	5	21	61			0.1	159	2			
COTTONSEED - WHOLE	8	12	38				160	0			
DDGS CORN	6	25	89	0.42	0.1	0.6	233	1			
DDGS CORN-WHEAT	34	70	114	0.56		0.4	275	1			
DDGS WHEAT	10	116	140	0.70		0.4	315	1			
DG WET WHEAT	51	100	140	0.70		0.4	315	1			
DISTILLERS G SYRUP	10	116	140	0.70		0.4	315	1			
GRAIN SCR PELLETS	6	17	40	0.03			133	1			
GREENFEED PELLETS	3	68	38	0.03			250	4			
HM CORN & COB	3										
LENTIL SCREENINGS	6	10	54	0.03			75	3			
LINSEED MEAL	29	42	36	0.91		0.2	354	2			
MOLASSES - WET	22	6	18			0.5	87	2			
MOLASSES - DRY	22	6	18			0.5	87	2			
PEA SCREENINGS	6	10	54	0.03			75	1			
POTATO WASTE - WET											
POTATOES	28	42	54	0.03			75	3			
SOYBEAN MEAL - 44%	22	35	57	0.51		0.1	185				
SOYBEAN MEAL - 48%	23	41	63	0.22	0.1	0.1	145	8			
SUNFLOWER SEEDS											
THIN STILLAGE	62	150	140	0.70			315	1			
TOMATO POMACE - DRY	33	47					900				
TURNIPS (TUBER)	21										
UREA											
WHEAT BRAN	10										
WHEAT SHORTS											9
WHOLE STILLAGE	40.00	125	60	0.65			315	1			

Table B3b: Typical nutrient analyses of western Canadian by-product samples.

Feed Name	DM %	DE Mcal/kg	CP %	NDF %	eNDF % of NDF	DIP % of CP	UIP % of CP	Ca %	P %	Mg %	K %	S %	Na %	Cl %
BARLEY GRAIN	88.5	3.66	12.5	23.0	34.0	67.0	33.0	0.07	0.38	0.14	0.54	0.14	0.02	
BARLEY GRAIN - HULLESS	88.5	3.78	12.2	23.0	34.0	67.0	33.0	0.06	0.36	0.14	0.54	0.14	0.02	
BARLEY GRAIN -HI MOIST	70.2	3.67	13.0	28.0	34.0	69.0	31.0	0.07	0.38	0.13	0.54	0.13	0.04	
CANOLA GRAIN	90.0	5.72	23.3	16.5	63.0	70.0	30.0	0.39	0.64	0.34	0.85	0.57	0.02	
CHICKPEA GRAIN	88.0	3.86	23.5	26.4	34.0	78.0	22.0	0.13	0.39	0.14	1.04	0.22	0.01	
CORN GRAIN	89.0	3.88	10.0	9.0	60.0	48.0	52.0	0.03	0.34	0.13	0.98	0.30	0.02	
FABABEANS	87.4	3.87	30.6	15.2	65.0	78.0	22.0	0.12	0.44	0.14	1.04	0.22	0.01	
FROZEN CANOLA	89.0	4.80	21.0	17.0	63.0	70.0	30.0	0.39	0.64	0.28	1.00	0.40	0.02	
HIGH MOISTURE CORN	88.0	3.96	10.8	10.0	60.0	44.7	55.3	0.03	0.29	0.11	6.00	0.60	1.48	
LENTIL SCREENINGS	88.2	3.30	21.2	14.0	20.0	78.0	22.0	0.31	0.45	0.14	1.04	0.22	0.01	
MIXED GRAIN	89.0	3.46	12.6	30.0	35.0	55.0	45.0	0.11	0.42	0.14	0.50	0.14	0.02	
OAT GRAIN	90.2	3.35	11.3	34.0	34.0	80.0	20.0	0.08	0.34	0.16	0.47	0.14	0.01	
PEA GRAIN	88.2	3.84	23.9	14.0	35.0	78.0	22.0	0.17	0.40	0.14	1.04	1.04	0.03	
PEA SCREENINGS	90.0	3.15	21.1	14.0	30.0	78.0	22.0	0.10	0.43	0.14	1.04	0.22	0.01	
RYE GRAIN	88.0	3.82	13.3	34.0	79.0	79.0	21.0	0.06	0.39	0.14	0.45	0.19	0.02	
SOYBEANS	90.0	4.15	38.0	14.9	30.0	75.0	25.0	0.28	0.65	0.23	2.01	0.35	0.04	0.03
TRITICALE GRAIN	90.2	3.69	16.1	19.0	34.0	79.0	21.0	0.06	0.34	0.17	0.49	0.13	0.02	
WHEAT GRAIN	88.0	3.86	14.2	11.7	84.0	74.0	26.0	0.05	0.41	0.16	0.02	0.14	0.03	

Table B4a: Typical nutrient analyses of Alberta grain samples.

Feed Name	DM %	DE Mcal/kg	CP %	NDF %	eNDF % of NDF	DIP % of CP	UIP % of CP	Ca %	P %	Mg %	K %	S %	Na %	Cl %
ALFALFA PAST - LV	24	2.73	19.7	56	56	81	19	1.96	0.3	0.17	2.09	0.17	0.02	
BLUEGRASS PAST - EB	20	3.04	16.6	50	98	79	21	0.46	0.21	0.11	1.2	0.11	0.1	
GRASS PAST - EV	23	3.26	21.3	42	41	94	6	0.42	0.37	0.19	2.5	0.15	0.02	
GRASS PAST - LV	25	2.95	13	23	90	90	10	0.41	0.2	0.15	1.4	0.16	0.03	
GRASS PAST - LB	24	2.86	12	42	92	85	15	0.53	0.22	0.17	1.3		0.03	
LEGUME PAST - EV	20	3.30	26	39	80	95	5	1.71	0.24	0.17	1.75	0.24	0.01	
LEGUME PAST - LV	24	2.86	17	42	80	95	5	1.86	0.19	0.17	2.14	0.24	0.03	
MIXED PAST - EV	21	3.30	18	45	95	80	20	1.07	0.21	0.16	1.5	0.17	0.03	
MIXED PAST - LV	22	2.73	13	52	95	80	20	1.03	0.17	0.16	1.75	0.18	0.02	
ORCHARDGR PAST - LV	24	2.82	17			100		0.42	0.19	0.16	1.25	0.11	0.01	
RED CLOVER PAST - MB	26	2.73	15.3			100		1.76	0.33	0.51	1.98			
REDTOP PAST - MB	39	2.60	7.4			100		0.33	0.23	0.18	2.13			
REED CAN PAST - MB	27	2.46	11.6			100		0.41	0.35	0.27	3.64			
TIMOTHY PAST - LV	26	2.68	18			100		0.39	0.32	0.15	2.4			
TREFOIL PAST - LV	24	2.77	21			100		1.91	0.22	0.28	1.99			

Table B5a: Typical nutrient analyses of Alberta pasture samples. EV - early vegetative; LV - late vegetative; EB - early bloom; MB - mid bloom; LB - late bloom.

Feed Name	Cu mg/kg	Mn mg/kg	Zn mg/kg	Se mg/kg	I mg/kg	Co mg/kg	Fe mg/kg	Mo mg/kg	Vit A KIU/kg	Vit D IU/kg	Vit E IU/kg
BARLEY GRAIN	6	17	40	0.03			80	2			
BARLEY GRAIN - HULLESS	6	17	40	0.03			80	2			
BARLEY GRAIN -HI MOIST	4	17	40	0.03			118	1			
CANOLA GRAIN	4	38	46	0.03			48	1			
CHICKPEA GRAIN	6	10	54	0.03			75	3			
CORN GRAIN	3	7	22	0.01			23	2			
FABABEANS	9	15	53	0.09			72	3			
FROZEN CANOLA	4	35	40	0.03			40	1			
HIGH MOISTURE CORN	22	6	18			0.5	87	1			
LENTILSCREENINGS	6	10	54	0.03			75	3			
MIXED GRAIN	3	15	35	0.03			70	1			
OAT GRAIN	4	39	28	0.03			74	2			
PEA GRAIN	6	10	54	0.03			75	3			
PEA SCREENINGS	6	10	54	0.03			152	1			
RYE GRAIN	6	34	48	0.03			156	1			
SOYBEANS	14	33	59	0.11			82				34
TRITICALE GRAIN	6	38	48	0.03			49	1			
WHEAT GRAIN	6	17	40				110	1			

Table B4b: Typical nutrient analyses of Alberta grain samples.

Feed Name	Cu mg/kg	Mn mg/kg	Zn mg/kg	Se mg/kg	I mg/kg	Co mg/kg	Fe mg/kg	Mo mg/kg	Vit A KIU/kg	Vit D IU/kg	Vit E IU/kg
ALFALFA PAST - LV	5	40	23	0.03			215	2			
BLUEGRASS PAST - EB	6	38	20	0.03			255	1			
GRASS PAST - EV	5	45	22	0.03			125	2			
GRASS PAST - LV	5	65	24	0.03			255	2			
GRASS PAST - LB	5	75	24	0.03			250	2			
LEGUME PAST - EV	6	40	23	0.03			213	2			
LEGUME PAST - LV	6	40	23	0.03			200	2			
MIXED PAST - EV	5	40	25	0.03			200	2			
MIXED PAST - LV	5	44	24	0.03			225	2			
ORCHARDGR PAST - LV	15										
RED CLOVER PAS T - MB	0										
REDTOP PAST - MB	0										
REED CAN PAST - MB	11										
TIMOTHY PAST - LV	3										
TREFOIL PAST - LV	7										

Table B5b: Typical nutrient analyses of Alberta pasture samples. EV - early vegetative; LV - late vegetative; EB - early bloom; MB - mid bloom; LB - late bloom.

Feed Name	DM %	DE Mcal/kg	CP %	NDF %	eNDF % of NDF	DIP % of CP	UIP % of CP	Ca %	P %	Mg %	K %	S %	Na %	Cl %
14% FINISHER RATION	90.0	1.29	15.5			70.0	30.0	0.88	0.44				0.39	
18% STARTER RATION	90.0	1.16	20.0			70.0	30.0	1.11	0.66				0.28	
32% LAMB SUPP 10% NPN	90.0	3.30	35.5			80.0	20.0	5.20	0.90	0.33	0.77	0.20	1.10	
32% LAMB SUPP ALL NAT	90.0	3.30	35.5			70.0	30.0	5.20	0.77	0.33	0.77	0.27	1.10	
36% SHEEP SUPP	90	3.30	40.0			70.0	30.0	3.90	0.89	1.11	1.89	0.44	0.93	
32% LAMB & SHEEP SUPP	90	3.30	35.6			70.0	30.0	3.10	1.22	0.56	1.11	0.78	1.11	
11/11 MINERAL PREMIX	99.0					70.0	30.0	11.0	11.0	0.40	1.00	0.10	11.00	
FEEDLOT MINERAL PREMIX	99.0					0.0		17.8	4.00	7.00	0.50	0.80	10.5	
20% MOLASSES BLOCK	70.0	3.30	26.6			70.0	30.0	6.66	2.66	4.00	2.66	0.80	4.00	

Table B6a: Typical nutrient analyses of commercial concentrate supplements offered for sale in western Canada.

Feed Name	Cu mg/kg	Mn mg/kg	Zn mg/kg	Se mg/kg	I mg/kg	Co mg/kg	Fe mg/kg	Mo mg/kg	Vit A KIU/kg	Vit D IU/kg	Vit E IU/kg
14% FINISHER RATION	5	40	25	0.33			15		13.7	1,777	28
18% STARTER RATION	5	40	25	0.33					13.3	2,666	55
32% LAMB SUPP 10% NPN	0	277	522	1.66	3.1	0.8	27		58.3	5,833	111
32% LAMB SUPP ALL NAT	0	277	522	1.66	3.1	0.8	27		58.3	5,833	111
36% SHEEP SUPP	16.7	394	478	2.22	8.9	3.3	257		41,111	6,889	289
32% LAMB & SHEEP SUPP	0	222	444		7.8	4.4	333		88,889	11,111	222
11/11 MINERAL PREMIX	0	4,970	8,800	30.0	125.0	38.0	4,500		1,000	100,000	2,000
FEEDLOT MINERAL PREMIX	0	1,680	2,800	24.0	60.0	25.0	1,585		500,000	50,000	2,500
20% MOLASSES BLOCK	40	2,000	2,400	8.00	53.3	13.3	1,333		13.3	40,000	2,666

Table B6b: Typical nutrient analyses of commercial concentrate supplements offered for sale in western Canada.

Feed Type		DM %	TDN %	CP %	Ca %	P %	K %	Mg %	Fe mg/kg	Mn mg/kg	Zn mg/kg	Cu mg/kg	Mo mg/kg
GRASS HAY	High	94.4	64.4	15.2	0.71	0.34	2.75	0.35	281	211	50	11.4	6.1
	Average	88.4	55.9	10.3	0.47	0.23	1.81	0.20	147	113	26	7.0	3.5
	Low	81.9	47.4	5.4	0.23	0.12	0.87	0.15	13	15	2	2.6	<1.0
GRASS LEGUME HAY	High	95.0	60.8	16.1	1.43	0.29	2.94	0.39	247	108	41	14.2	5.6
	Average	87.3	54.2	11.8	0.87	0.22	2.06	0.24	133	59	24	8.7	3.1
	Low	79.6	47.7	7.5	0.31	0.15	1.18	0.09	19	9	7	3.2	<1.0
ALFALFA HAY	High	95.9	64.9	19.4	1.93	0.37	3.21	0.43	446	58	38	15.1	4.3
	Average	88.7	57.0	16.4	1.33	0.27	2.47	0.28	251	35	23	9.7	2.6
	Low	81.5	49.2	13.4	0.73	0.17	1.73	0.13	56	13	9	4.2	<1.0
CEREAL HAY	High	95.2	64.9	14.0	0.57	0.37	3.04	0.33	360	100	40	14.6	4.1
	Average	86.4	60.9	9.3	0.37	0.26	1.95	0.18	194	55	26	7.8	2.4
	Low	77.6	56.9	4.6	0.17	0.17	0.86	0.03	28	10	12	1.0	<1.0
BARLEY GRAIN	High	92.6	84.6	13.2	0.19	0.46	0.63	0.17	19	36	59	17.8	2.3
	Average	88.6	81.4	11.2	0.11	0.38	0.53	0.15	119	22	45	11.0	1.7
	Low	84.5	78.2	9.3	0.03	0.30	0.43	0.13	219	8	32	4.2	<1.0
OATS GRAIN	High	90.8	80.6	12.8	0.18	0.40	0.78	0.15	121	70	45	14.6	3.8
	Average	87.7	76.9	10.9	0.10	0.34	0.70	0.13	79	48	37	7.8	2.4
	Low	84.5	73.2	9.0	0.02	0.28	0.62	0.11	38	26	28	1.0	<1.0
GRASS SILAGE	High	47.5	59.7	16.0	0.82	0.60	3.43	0.42	570	198	55	15.8	4.4
	Average	35.1	53.3	12.6	0.57	0.40	2.47	0.30	352	124	35	9.1	2.6
	Low	22.7	47.0	9.3	0.32	0.20	1.51	0.18	134	50	15	2.4	0.8
GRASS LEGUME SILAGE	High	46.6	58.1	18.7	1.33	0.34	2.98	0.38	509	122	51	13.9	5.6
	Average	34.4	53.3	15.5	0.93	0.26	2.28	0.25	287	76	31	8.7	3.2
	Low	22.2	48.5	12.2	0.53	0.18	1.58	0.12	65	30	11	3.5	0.7
CORN SILAGE	High	37.6	71.8	10.8	0.60	0.30	1.63	0.32	360	80	47	11.2	2.4
	Average	29.6	63.9	8.8	0.36	0.24	1.29	0.20	213	47	29	7.0	1.7
	Low	21.6	56.0	6.8	0.12	0.18	0.95	0.08	66	14	11	2.8	1.0
CEREAL SILAGE	High	53.4	77.9	11.6	0.60	0.40	2.38	0.27	498	113	48	13.1	3.9
	Average	38.2	63.4	9.1	0.39	0.28	1.61	0.19	283	66	32	7.1	2.5
	Low	22.9	48.9	6.7	0.18	0.16	0.84	0.11	68	18	16	1.1	1.1

Table B7: Summary of analyses of feed samples submitted to the BC Soil and Feed Test Laboratory, 1969-1984.

Feed Name	DM %	DE Mcal/kg	CP %	ADF %	NDF %	RFV %	Ca %	P %	K %	Mg %	S %	Na %	Cu mg/kg	Fe mg/kg	Mn mg/kg	Mo mg/kg	Zn mg/kg
ALFALFA	83.2	2.69	20.3	34.2	45.5	134	1.59	0.23	2.24	0.35	0.03	0.05	7.55	164	39.9	3.26	19.1
ALFALFA/GRASS	83.5	2.58	16.1	16.5	53.0	109	1.19	0.20	1.99	0.30		0.03	6.11	116	36.4	2.91	18.3
ANNUAL RYEGRASS	78.7	2.56	11.6	38.2	62.7	88	0.46	0.26	1.78								
CRESTED WHEATGRASS	85.4	2.57	9.3	37.5	62.7	90	0.29	0.15	1.48	0.14		0.02					
MEADOW BROME GRASS	83.4	2.47	10.8	41.2	65.3	84	0.51	0.15	1.93	0.19		0.02	2.4	54	21	5.1	13.6
ORCHARDGRASS	80.8	2.61	14.2	38.1	64.7	89	0.47	0.19	1.96	0.28		0.09	5.1	64	29.4	1	14.9
PERENNIAL RYE GRASS	86.0	2.44	8.5	40.6	65.6	82	0.33	0.15	1.42	0.21		0.03					
REED CANARY GRASS	88.2	2.62	15.2	37.5	67.7	86	0.34	0.22	2.29								
RUSSIAN WILD RYE GRASS	82.3	2.53	14.9	40.0	67.5	83	0.39	0.21	2.19								
SMOOTH BROME GRASS	86.0	2.54	12.1	39.5	65.8	85	0.28	0.23	2.85								
TALL FESCUE	83.0	2.59	11.6	39.7	68.1	83	0.37	0.12	1.97	0.26		0.04					
TIMOTHY	85.6	2.50	8.2	40.6	65.0	83	0.33	0.16	1.47	0.14		0.01	2.05	53	32.3	2.46	13.5
GRASS/ALFALFA	83.6	2.61	13.2	37.8	56.9	98	0.81	0.19	1.93	0.24		0.02	4.86	107	32.2	2.67	16.5
CLOVER	85.4	2.61	16.5	36.5	49.2	118	1.28	0.23	2.27								
CLOVER/GRASS	83.7	2.48	14.8	40.0	52.7	105	1.01	0.18	1.99	0.35		0.02	6.4	106	48.4	5.2	4.8
QUACK GRASS	81.4	2.54	8.9	39.0	63.0	89	0.53	0.17	0.62	0.13		0.02					
KOCHIA	80.5	2.46	12.7	40.2	74.1	58	0.91	0.20	1.61	0.65		1.32					
NATIVE	82.3	2.47	10.0	40.6	61.4	89	0.47	0.11	1.11	0.25		0.05	2.6	136	97.8	0.72	22.8
SLOUGH	84.3	2.45	9.5	41.2	66.5	79	0.54	0.12	1.30	0.27		0.08	3.5	199	110	1.4	15.9

Table B8: Summary of analyses of hay samples submitted by Manitoba livestock producers, 2000-2005.

Feed Name	DM %	DE Mcal/kg	CP %	ADF %	NDF %	RFV %	Ca %	P %	K %	Mg %	S %	Na %	Cu mg/kg	Fe mg/kg	Mn mg/kg	Mo mg/kg	Zn mg/kg
ALFALFA	46.1	2.63	19	36.7	45.8	129	1.54	0.26	2.44	0.38	0.27	0.05	6.5	206	36	3.19	24.1
ALFALFA/GRASS	67.3	2.62	17.4	36.7	50.2	115	1.41	0.25	2.36	0.35		0.07	8.02	118	37.3	1.96	22.6
BARLEY	40.9	2.89	11.1	33.1	53.3	110	0.43	0.26	1.71			0.19	4.93	224	29.2	1.65	25.2
CANOLA	34.4	2.64	14.7	37.5	46.3	122	1.64	0.28	1.71		0.64	0.13	4.6	119	35.9	1	21.7
CORN	34.7	2.94	9.4	30.1	53.6	116	0.28	0.24	1.19			0.03	3.92	192	28.3	1.19	24.8
FALL RYE	37.1	2.76	11.9	37.7	61.1	87	0.43	0.28	1.74			0.04					
MILLET	42.3	2.64	11.9	36.8	62.3	90	0.43	0.24	2.52			0.03					
OATS	38.7	2.77	10.9	37	58.8	94	0.36	0.25	2.02			0.3	4.25	196	38.2	2.67	21.8
SORGHUM/SUDAN GRASS	37.6	2.65	11.9	39.3	62.4	86	0.69	0.23	2.42		0.15	0.05	1.9	140	4.5	1.4	10.2
WHEAT	50.9	2.86	10.1	34.6	56.5	95	0.23	0.23	1.74			0.05					
BARLEY + OATS	39.2	2.80	13.3	35.4	55.7	103	0.42	0.22	1.85	0.35		0.29	6.3	107	17.7	1.6	20.6
BARLEY + PEAS	37.8	2.88	13.5	32.9	50.3	116	0.68	0.3	1.55	0.35		0.17					

Table B9: Summary of analyses of silage samples submitted by Manitoba livestock producers, 2000-2005.

Feed Name	DM %	DE Mcal/kg	CP %	ADF %	NDF %	RFV %	Ca %	P %	K %	Mg %	S %	Na %	Cu mg/kg	Fe mg/kg	Mn mg/kg	Mo mg/kg	Zn mg/kg
BARLEY	74.0	2.74	11.4	35.2	56.4	102	0.45	0.25	1.78	0.25		0.17	4.32	190	25.4	2.65	19.4
CANOLA	73.4	2.48	12.3	39.0	52.1	105	1.52	0.27	1.43	0.35	0.59	0.07	4.67	116	30.9	2.03	18.9
CORN	34.0	2.91	9.8	32.6	51.9	128	0.23	0.27	1.15	0.24		0.01	3.04	250	20.7	1.67	19.5
FALL RYE	79.7	2.48	12.7	39.3	58.8	99	0.40	0.25	1.78	0.18		0.07	0.40				
MILLET	75.8	2.63	10.9	36.9	61.5	92	0.44	0.22	2.20	0.39		0.05	8.30	409	29.7	1.40	44.0
OATS	76.6	2.65	10.3	37.0	58.7	96	0.33	0.23	2.03	0.22	0.26	0.30	3.80	114	32.4	1.80	18.1
SORGHUM/SUDAN GRASS	63.2	2.65	11.5	37.3	62.7	89	0.50	0.21	2.08	0.41		0.09	0.50				
WHEAT	74.8	2.52	10.7	40.2	60.8	89	0.24	0.23	1.42	0.16		0.05	2.60	93	57.2	0.80	18.9

Table B10: Summary of analyses of 'greenfeed' samples submitted by Manitoba livestock producers, 2000-2005. Note that in Manitoba, 'greenfeed' refers to a crop fed fresh, either directly after cutting or after a short period of wilting after cutting. In BC, this type of feed is often called 'green chop'. In Alberta, 'greenfeed' commonly refers to a grain crop fed as hay.

Feed Name	DM %	DE Mcal/kg	CP %	ADF %	NDF %	RFV %	Ca %	P %	K %	Mg %	S %	Na %	Cu mg/kg	Fe mg/kg	Mn mg/kg	Mo mg/kg	Zn mg/kg
BARLEY	80.6	1.99	5.5	49.4	74.1	63	0.36	0.09	1.76	0.17		0.24	2.92	127	22.4	2.98	12.7
CORN	68.7	2.50	5.7	45.6	76.2	64	0.32	0.17	1.42	0.31		0.02					
FLAX	85.4	1.55	6.7	53.1			0.61	0.11	0.71	0.33		0.10					
OATS	80.5	2.05	5.3	49.4	73.0	64	0.29	0.10	2.32	0.16	0.16	0.31	3.11	86	19.5	2.08	11.4
PEA	83.4	2.04	6.5	48.3	65.8	72	1.65	0.09	1.48	0.44		0.08					
WHEAT	82.7	1.89	4.4	51.4	76.2	60	0.22	0.08	1.39	0.14		0.06	3.48	47	23.9	2.20	11.6

Table B11: Summary of analyses of straw samples submitted by Manitoba livestock producers, 2000-2005.

Feed Name	DM %	DE Mcal/kg	CP %	ADF %	NDF %	Ca %	P %	K %	Mg %	S %	Na %	Cu mg/kg	Fe mg/kg	Mn mg/kg	Mo mg/kg	Zn mg/kg
BARLEY	86.5	3.69	12.6	7.5	21.2	0.06	0.37	0.53	4.13	72	0.14	19.7	2.53	0.02	0.14	32
OATS	87.7	3.30	12.4	15.5	31.8	0.10	0.37	0.46			0.16			0.02		

Table B12: Summary of analyses of grain samples submitted by Manitoba livestock producers, 2000-2005.

Appendix C: Example Rations

The rations illustrated on the next few pages are intended only to demonstrate a typical format for ration formulation. For these, an Excel® workbook was used, taking advantage of Excel's Solver® add-in. To obtain a copy of the wokbook, contact steve@agromedia.ca.

Sheep Ration Formulator

MEWL-MGP-30 Mixed Gender Early Weaned Lambs: Moderate Growth Potential (Medium-Sized Ram Growth Standard)

Body Weight 30 kg Litter or Milk Weight 0 kg ADG 400 g/day

Diet Ingredients	DM Inclusion Rate, % Min	Max	Solution	DMI kg	As-fed kg	DM %	DE Mcal/kg	CP %	NDF %	Ca %	P %	A-F Cost $/tonne
Barley Silage	0	100	21.0	0.273	0.741	36.8	2.75	11.1	49.0	0.46	0.27	45.00
Barley Grain	0	100	68.2	0.886	1.001	88.5	3.66	12.5	23.0	0.07	0.38	196.00
32% Lamb Supp All Nat CP	0	100	10.8	0.141	0.157	90.0	3.30	35.5	0.0	5.20	0.77	654.20
Requirement			100.0	1.3			3.43	14.7	20.0	0.51	0.24	
Maximum			100.0	1.3			3.50	17.0	33.0	0.60	0.40	
Solution			100.0	1.3	1.9	68.4	3.43	14.7	26.0	0.71	0.40	174.85
Excess/(Deficit)			0.0	0.0			0.00	(0.0)	6.0	0.20	0.16	

Min-Vit Supp Requirement 0.02 0.00 0.00

Diet Ingredients	Na %	Cl %	K %	Mg %	S %	Co	Cu	Fe	I	Mn mg/kg	Mo	Se	Zn	vit A IU/kg	vit E IU/kg
Barley Silage	0.20	0.00	1.60	0.22	0.22	0.00	5.0	300	0.0	33.0	2.0	0.030	27	0	0
Barley Grain	0.02	0.00	0.54	0.14	0.14	0.00	6.0	80	0.0	17.0	0.0	0.030	40	0	0
32% Lamb Supp All Nat CP	1.10	0.00	0.77	0.33	0.27	0.80	0.0	27	3.1	277.0	0.0	1.660	522	58	111
Requirement	0.06	0.05	0.48	0.11	0.18	0.15	7.50	89.00	0.50	19.00	0.50	0.340	7	1,085	15
Maximum						10	25	500	50	1000	10	2	750		
Solution	0.17	0.00	0.79	0.18	0.17	0.09	5.14	120.42	0.34	48.55	0.42	0.207	90	6	12
Excess/(Deficit)	0.11	(0.05)	0.31	0.07	(0.01)	(0.06)	(2.36)	31.4	(0.2)	29.5	(0.1)	(0.133)	82	-1,079	-3
Min-Vit Supp Requirement	0.00	2.50	0.00	0.00	0.46	3.16	118.02	0.0	8.2	0.0	4.0	6.663	0	53,934	148

Figure C1: Example ration formulated for mixed gender early weaned lambs with moderate growth potential with reference to the medium-sized ram growth standard shown in figure A1.

Sheep Ration Formulator

EL-ELT-40   Ewe Lambs: First 6-8 weeks of lactation suckling twins (wean by 8 weeks)

Body Weight 40 kg   Litter or Milk Weight 1 kg   ADG -100 g/day

Diet Ingredients	DM Inclusion Rate, % Min	DM Inclusion Rate, % Max	Solution	DMI kg	As-fed kg	DM %	DE Mcal/kg	CP %	NDF %	Ca %	P %	A-F Cost $/tonne
Alfalfa-Grass Hay	0	100	42.7	0.897	1.026	87.4	2.63	14.0	52.0	1.22	0.19	165.00
Oat Grain	0	100	52.2	1.095	1.214	90.2	3.35	11.3	34.0	0.08	0.34	184.00
CJ 32% Lamb & Sheep Supp	0	100	5.1	0.108	0.120	90.0	3.30	35.6	0.0	3.11	1.22	285.00
Requirement			100.0	2.1			3.04	13.7	20.0	0.37	0.26	
Maximum							3.50	17.0	33.0	0.60	0.40	
Solution			100.0	2.1	2.4	89.0	3.04	13.7	39.9	0.72	0.32	180.88
Excess/(Deficit)			0.0	0.0	0.0		0.00	(0.0)	19.9	0.35	0.06	
Min-Vit Supp Requirement				0.02						0.00	0.00	

Diet Ingredients	Na %	Cl %	K %	Mg %	S %	Co mg/kg	Cu mg/kg	Fe mg/kg	I mg/kg	Mn mg/kg	Mo mg/kg	Se mg/kg	Zn mg/kg	vit A IU/kg	vit E IU/kg
Alfalfa-Grass Hay	0.01	0.00	1.65	0.26	0.17	0.00	4.0	200	0.0	40.0	2.0	0.030	23	0	0
Oat Grain	0.01	0.00	0.47	0.16	0.14	0.00	4.0	74	0.0	39.0	0.0	0.030	28	0	0
CJ 32% Lamb & Sheep Supp	1.11	0.00	1.11	0.56	0.78	4.44	0.0	333	7.8	222.2	0.0	0.000	444	88,889	222
Requirement	0.04	0.03	0.48	0.04	0.18	0.15	2.70	7.00	0.80	8.60	0.50	0.157	33	2,292	15
Maximum						10	25	500	50	1000	10	2	750		
Solution	0.07	0.00	1.01	0.22	0.19	0.23	3.79	141.13	0.40	48.85	0.85	0.028	47	4,570	11
Excess/(Deficit)	0.03	(0.03)	0.53	0.18	0.01	0.08	1.09	134.1	(0.4)	40.2	0.4	(0.129)	14	2,278	-4
Min-Vit Supp Requirement	0.00	1.50	0.00	0.00	0.00	0.00	0.00	0.0	20.0	0.0	0.0	6.427	0	0	179

Figure C2: Example ration formulated for ewe lambs during first 6-8 weeks of lactation suckling twins.

Sheep Ration Formulator

YME-ELT-70 Yearling And Mature Ewes: First 6-8 weeks of lactation suckling twins

	Body Weight	70	kg	Litter or Milk Weight	1.7	kg	ADG	-60	g/day

Diet Ingredients	DM Inclusion Rate, % Min	Max	Solution	DMI kg	As-fed kg	DM %	DE Mcal/kg	CP %	NDF %	Ca %	P %	A-F Cost $/tonne
Grass Haylage	0	100	59.3	1.661	3.690	45.0	2.46	12.2	62.0	0.90	0.22	165.00
Barley Grain	0	100	29.3	0.819	0.926	88.5	3.66	12.5	23.0	0.07	0.38	184.00
Canola Meal	0	100	10.7	0.300	0.327	91.9	3.08	38.3	27.0	0.75	1.26	285.00
Drylot Mineral Premix	0	100	0.7	0.020	0.020	99.0	0.00	0.0	0.0	17.80	4.00	654.20
Requirement			100.0	2.8			2.86	15.0	20.0	0.39	0.29	
Maximum							3.50	17.0	33.0	0.60	0.40	
Solution			100.0	2.8	5.0	56.4	2.86	15.0	46.4	0.76	0.41	178.44
Excess/(Deficit)			0.0	0.0			(0.00)	(0.0)	26.4	0.37	0.12	
Min-Vit Supp Requirement				0.02						0.00	0.00	

Diet Ingredients	Na %	Cl %	K %	Mg %	S %	Co mg/kg	Cu mg/kg	Fe mg/kg	I mg/kg	Mn mg/kg	Mo mg/kg	Se mg/kg	Zn mg/kg	vit A IU/kg	vit E IU/kg
Grass Haylage	0.03	0.00	2.27	0.23	0.18	0.00	5.0	125	0.0	65.0	1.0	0.030	26	0	0
Barley Grain	0.02	0.00	0.54	0.14	0.14	0.00	6.0	80	0.0	17.0	0.0	0.030	40	0	0
Canola Meal	0.03	0.00	1.31	0.62	1.16	0.00	9.0	200	0.0	58.0	0.0	0.600	97	0	0
Drylot Mineral Premix	10.50	0.00	0.50	7.00	0.80	25.00	0.0	1585	60.0	1,680.0	0.0	24.000	2,800	500,000	2,500
Requirement	0.06	0.10	0.53	0.10	0.18	0.15	3.20	9.00	0.80	11.10	0.50	0.201	39	2,500	15
Maximum						10	25	500	50	1000	10	2	750		
Solution	0.10	0.00	1.65	0.29	0.28	0.18	5.69	130.31	0.43	61.74	0.59	0.262	58	3,571	18
Excess/(Deficit)	0.04	(0.10)	1.12	0.19	0.10	0.03	2.49	121.3	(0.4)	50.6	0.1	0.061	19	1,071	3
Min-Vit Supp Requirement	0.00	5.00	0.00	0.00	0.00	0.00	0.00	0.00	18.6	0.0	0.0	0.000	0	0	0

Figure C3: Example ration formulated for Eewes during first 6-8 weeks of lactation suckling twins.

Appendix D: Energy Conversions

TDN %	DE Mcal/kg	ME Mcal/kg	NEm Mcal/kg	NEg Mcal/kg	TDN %	DE Mcal/kg	ME Mcal/kg	NEm Mcal/kg	NEg Mcal/kg
20	0.88	0.72	-0.20	-0.71	60	2.64	2.16	1.30	0.74
21	0.92	0.76	-0.16	-0.67	61	2.68	2.20	1.33	0.77
22	0.97	0.79	-0.12	-0.62	62	2.73	2.23	1.37	0.80
23	1.01	0.83	-0.07	-0.58	63	2.77	2.27	1.40	0.83
24	1.06	0.86	-0.03	-0.54	64	2.82	2.30	1.43	0.86
25	1.10	0.90	0.01	-0.50	65	2.86	2.34	1.46	0.89
26	1.14	0.94	0.05	-0.46	66	2.90	2.38	1.50	0.92
27	1.19	0.97	0.09	-0.42	67	2.95	2.41	1.53	0.95
28	1.23	1.01	0.13	-0.38	68	2.99	2.45	1.56	0.98
29	1.28	1.04	0.17	-0.34	69	3.04	2.48	1.59	1.01
30	1.32	1.08	0.21	-0.30	70	3.08	2.52	1.62	1.04
31	1.36	1.12	0.25	-0.26	71	3.12	2.56	1.66	1.06
32	1.41	1.15	0.29	-0.22	72	3.17	2.59	1.69	1.09
33	1.45	1.19	0.33	-0.18	73	3.21	2.63	1.72	1.12
34	1.50	1.22	0.37	-0.14	74	3.26	2.66	1.75	1.15
35	1.54	1.26	0.41	-0.10	75	3.30	2.70	1.78	1.17
36	1.58	1.30	0.45	-0.07	76	3.34	2.74	1.81	1.20
37	1.63	1.33	0.48	-0.03	77	3.39	2.77	1.84	1.23
38	1.67	1.37	0.52	0.01	78	3.43	2.81	1.87	1.26
39	1.72	1.40	0.56	0.04	79	3.48	2.84	1.90	1.28
40	1.76	1.44	0.60	0.08	80	3.52	2.88	1.93	1.31
41	1.80	1.48	0.64	0.12	81	3.56	2.92	1.96	1.33
42	1.85	1.51	0.67	0.15	82	3.61	2.95	1.99	1.36
43	1.89	1.55	0.71	0.19	83	3.65	2.99	2.02	1.39
44	1.94	1.58	0.75	0.22	84	3.70	3.02	2.05	1.41
45	1.98	1.62	0.78	0.26	85	3.74	3.06	2.08	1.44
46	2.02	1.66	0.82	0.29	86	3.78	3.10	2.11	1.46
47	2.07	1.69	0.85	0.33	87	3.83	3.13	2.14	1.49
48	2.11	1.73	0.89	0.36	88	3.87	3.17	2.17	1.51
49	2.16	1.76	0.92	0.39	89	3.92	3.20	2.20	1.54
50	2.20	1.80	0.96	0.43	90	3.96	3.24	2.23	1.56
51	2.24	1.84	1.00	0.46	91	4.00	3.28	2.26	1.59
52	2.29	1.87	1.03	0.49	92	4.05	3.31	2.29	1.61
53	2.33	1.91	1.06	0.52	93	4.09	3.35	2.31	1.64
54	2.38	1.94	1.10	0.56	94	4.14	3.38	2.34	1.66
55	2.42	1.98	1.13	0.59	95	4.18	3.42	2.37	1.68
56	2.46	2.02	1.17	0.62	96	4.22	3.46	2.40	1.71
57	2.51	2.05	1.20	0.65	97	4.27	3.49	2.43	1.73
58	2.55	2.09	1.23	0.68	98	4.31	3.53	2.46	1.75
59	2.60	2.12	1.27	0.71	99	4.36	3.56	2.49	1.78

Table C1: Equivalent energy values calculated using the following equations:
DE Mcal/kg = TDN % x 0.044; ME Mcal/kg = TDN % x 0.036;
$NEm = 1.37 x ME - 0.138 x ME^2 + 0.0105 x ME^3 - 1.12$;
$NEg = 1.427 x ME - 0.174 x ME^2 + 0.0122 x ME^3 - 1.65$

Appendix E: Sources

NRC. 1985. Nutrient Requirements of Sheep, 6th rev. ed. National Academy Press, Washington, DC.

NRC. 2007. Nutrient Requirements of Small Ruminants: sheep, goats, cervids, and New World camelids, National Academy Press, Washington, DC.

Tedeschi, L.O. and Fox D.G. 2018. The Ruminant Nutrition System: An Applied Model for Predicting Nutrient Requirements and Feed Utilization In Ruminants, 2nd edition. XanEdu, Acton MA.

AFRC (Agriculture and Food Research Council). 1993. Energy and Protein Requirements of Ruminants. CAB International, Wallingford, UK.

Cannas, A., Tedeschi, L.O., Fox, D.G., Pell, A.N. and Van Soest, P.J. 2004. A mechanistic model for predicting the nutrient requirements and feed biological values for sheep. J. Anim. Sci. 82: 149-169.

Cannas, A., Tedeschi, L.O., Atzori, A.S. and Fox, D.G. 2010. The development and evaluation of the Small Ruminant Nutrition System. p 263-272 in D. Sauvant et al. (eds.), Modelling nutrient digestion and utilisation in farm animals, Wageningen Academic Publishers, Wageningen NL.

CSIRO (Australian Commonwealth Scientific and Industrial Research Organization). 2007. Nutrient Requirements of Domesticated Ruminants. CSIRO Collingwood, Australia.

Index

Symbols

www.ingramcontent.com/pod-product-compliance
Lightning Source LLC
Chambersburg PA
CBHW041607220326

41597CB00052B/66